Mobile Web 2.0

The innovator's guide to developing
and marketing next generation
wireless/mobile applications

By
Ajit Jaokar and Tony Fish

Copyright 2006 Futuretext Limited

Issue Date 15 Aug 2006
Published by
Futuretext
36 St George St
Mayfair
London
W1S 2FW, UK
Email:info@futuretext.com
www.futuretext.com

ISBN: 0-9544327-6-2

Contents

          Ajit Jaokar and Tony Fish

         Ajit Jaokar and Tony Fish

　　　　　　　　　　Ajit Jaokar and Tony Fish

Acknowledgements

"Mobile Web 2.0" the book was born in October 2005. It has been a significant undertaking, much more than we first thought; primarily because we have had to generate the ideas, test them via our networks and blogs and then refine the concepts. This took a lot of time, especially because we had to intellectually engage with many of the people worldwide who responded to our posts.

Thus, there is no doubt that we could not have written this book without help from so many people within our trusted network, friends and other Web 2.0 enthusiasts.

Journalist Mike Butcher (who writes for the FT as well as mobbites.com and tbites.com) acted as both a consulting editor on the book and a foil for some of the ideas expressed. Many thanks to Mike for his editing and feedback.

Also, the book would not have been possible without the untiring efforts of Maggie Baldry who edited and proof read the book. We would like to express special thanks to Maggie for helping us with the very tight deadlines

We would also like to thank members of Forum Oxford and the Ecademy Mobile Applications Club for the feedback to our ideas and for the input from the Web 2.0 workgroup.

Ajit would like to thank Aditya (for permitting me to use his 'gagga' i.e. computer), Geetha (who looks forward to a world tour), his parents and Dr Peter Gray for their support.

Tony wishes to say a special thanks to his enduring wife, Nicky, who is willing to put up with the stupid hours and never ending phone calls. And to his growing daughters Ellie and Millie who are definitely part of the next generation of mobile users.

Significant contributions have been made by Bill Jones of Global Village, Alan Patrick of Broadsight, Aenil Premji of AMF, Sam Sethi of Vecosys and Philip Otley of Accenture.

We also need to express our thanks to Phil Hutton, Tomi Ahonen, Dr Rebecca Lingwood, Dion Hinchcliffe, Roger Strukhoff, Jeremy Geelan, Richard MacManus, Ian Tomson-Smith, Margaret Gold and Rudy De Waele

How To Read This Book

This book has three parts. Part One discussses Web 2.0 and Mobile Web 2.0 in detail. Part Two contains various factors that influence Mobile Web 2.0. Finally, Part Three outlines business models, funding etc.

If the style sounds informal and conversational; it is! The book was blogged extensively on OpenGardens blog[1] and Forum Oxford[2]. Proofreaders and Editors have combed the book extensively, but we apologise for any gramattical errors that have crept in.

---

1 www.opengardensblog.futuretext.com
2 www.forumoxford.com

Ajit Jaokar and Tony Fish

Foreword

So, what happened to Mobile Web 1.0? For those who missed this experience, it felt a bit like how the first TV viewers saw "radio with pictures". Despite a number of creative attempts to break the model - of wireless companies pushing a new service to their captive base - the underwhelming take-up rates for the Mobile Web reflected a missed target. For many reasons, good and bad, the creative jungle of the Web that had thrown up blogging, contextual search, various "MySpaces" and a host of other compelling user propositions, had failed to transfer to the mobile arena.

The authors of Mobile Web 2.0 assert that "Mobile" can be a major factor in the changing balance of power in the converging digital world. They assert that broadcast content as we know it is losing its appeal, that the walled gardens set up to protect revenue streams from the free for all (in every sense) of the Web are collapsing. This is principally being brought about by the impact of user generated content, captured in context as only a mobile device can do. Mobile Web 2.0 is focused on the user as the creator and consumer of content 'at the point of inspiration' and the mobile device as the means to harness collective intelligence.

Mobile Web 2.0 (the book) is not about repeating the hype of Mobile Web 1.0. It is about stimulating debate and explaining the possible scenarios that the inevitable change will bring. The underlying technology is clearly presented, and the concepts explained creatively. The most heated debate will probably be around the potential impact the Mobile Web may have on the worlds of mobile, media and the Web.

Building on the underlying belief that a mobile device will bring about significant change to the communications, media, and web industries; as their business models move from supply-demand into create-consume, Mobile

Web 2.0 re-defines success factors for executives, VCs or developers seeking to take advantage of this shift. To thrive, many companies must master the art of the "ring master" rather than that of the "director" of their sectors, skilfully adapting to the performances of a growing and potentially huge range of performers.

These capabilities are related but very different - those who seek to create rules without reason to govern this circus will likely be the losers. Many of the old rules will still apply – timing of your own moves as the critical factor being foremost amongst them. Some readers may struggle with the concepts put forward, as many will find it hard to struggle with innovation outside their own companies. Nonetheless, the authors have set out a clear case for change, and that alone should bring about discussion that is more informed and a raised awareness of a rapidly changing industry structure. Let the debate begin…

Philip Otley
Partner, Wireless Industry, Accenture

Ajit Jaokar and Tony Fish

Introduction

Background

This book holds that the mobile phone is the pivotal device for the next stage of web development, but on radically different terms when compared to today's regimented environment of 'push' content. Every one of us is both a potential creator and a consumer of content in a number of forms: from our simple diary dates right up to complex opinions and thoughts. In the modern world, there is now an inbuilt creative / consumption balance in all aspects of our life spread across entertainment, web and communications. Far from being a mere passive consumer of content as it is now, the mobile device is poised to be a significant tool in the creation of content. It is optimally and uniquely placed to capture content at the 'point of inspiration'. This makes it central to the next incarnation of the Web and what has been known as Web 2.0.

Web 2.0 could be described as the **intelligent web** or **harnessing collective intelligence**. The rise of "User Generated Content" online, where users are given control of the content creation process, is creating a second wave in the digital media field. Coupled with the growing technical capacity of mobile phones, this phenomenon is very significant. Historically, the mobile device has always been the mechanism to capture content at the point of inspiration. Unlike a camera, a voice recorder or a PC, the mobile device is always with us. But with Web 2.0, for the first time, we have the capacity to add intelligence to the content we capture at the point of inspiration (via mechanisms like tags) and also a method to glean intelligence from the aggregation of tagged content. This is achieved through contextual 'tags', an on-the-fly taxonomy created by the user, also known as 'folksonomy'.

This idea of harnessing collective intelligence through mobile devices which can capture content at the point of inspiration is what we call 'Mobile Web 2.0'. This model is one of Destructive Creativity.

This book looks at mobile applications in an unconventional manner. We don't talk much about the traditional media applications; but rather we focus on the new applications driven by user generated content. Your view of mobile applications and the Web will directly influence your value view of this book.

- If you see mobile applications as purely an extension of the Web, you are going to struggle with the concepts and business models

- If you see mobile applications as merely an access technology within a stand alone industry, protected by airtime licenses and technology; put this book back on the shelf and move on

- If you see that mobile applications and the Web are symbiotic with distinct and unique values that need to learn and grow together; you are not going to put this down..

We will be talking a lot about cross sector applications and convergence. True convergence centres on people. It is all about the users' ability to share their views, their opinions, their dates, their events, their joys and their traumas and to be entertained and informed by other individuals' day to day life experiences; all without the intermediary and biased media circus. This will be a fundamental and seismic shift in business models especially for media companies. Further, it could be argued that Web 2.0 is not about invention of new technologies, but rather the exploitation of existing technologies for the re-invention of the media industry.

The first edition of Mobile Web 2.0 was called 'OpenGardens'. OpenGardens was a phrase we coined to connote the opposite of the 'walled gardens' in the Mobile Data Industry. In that book, published in late 2004, we asked:

- Why is the Mobile Web different from the traditional Web? Why can't they be seamless?

Ajit Jaokar and Tony Fish

- Why do so many application developers struggle to make a profit in the Mobile Data Industry?

- Why do we still see the 'walled gardens' on the Mobile Internet?

- Why do we see so many entertainment type applications (ringtones, pictures etc) as opposed to (useful) utility type applications ('find my nearest' services)?

- What factors are hampering critical mass (network effect) on the Mobile Data Industry?

While originally designed for voice, mobile phones are also capable of running applications handling data. This gave rise to a whole new industry i.e. the Mobile Data Industry. Knowingly or unknowingly, most of us are familiar with the Mobile Data Industry. On a daily basis, over 2 billion of us interact with it. A majority of the population in Western Europe, USA and Asian countries possess a mobile device of some kind. The phone was sold for and is used for making phone calls (Voice traffic). In addition, a majority of users use the text messaging service leading to SMS (Short Messaging Service) traffic. This 'data' service is often a secondary use for the phone in comparison to voice. Even though there are over 5,000 other mobile applications, SMS is the first introduction to mobile data for most people. Sadly, often that's far as they get in their usage of mobile data.

The Mobile Data Industry (which commenced around 1996) is closely related to the rise of the World Wide Web. The Mobile Data Industry is all around you - walk into most public places and you can't fail to be distracted by the chimes of ringtones for instance. In addition, there are the other devices – such as PDA (Personal Digital Assistants), RIM blackberry etc. all busy sending and receiving data. Over the last few years, the growth of Mobile Data (as opposed to voice) has shown dramatic increases, now accounting for up to 25% of a Mobile operator's revenues in some cases. The industry is growing at a scorching pace. According to IDC (www.idc. com), during the second quarter of 2005, handset shipments jumped 7.3 percent over the previous quarter and 16.3 percent to the previous year to 188.7 m units globally. A majority of these new handsets now handle both voice and data ARPU (Average revenue per user) keeps increasing steadily as we can see in figure 1.

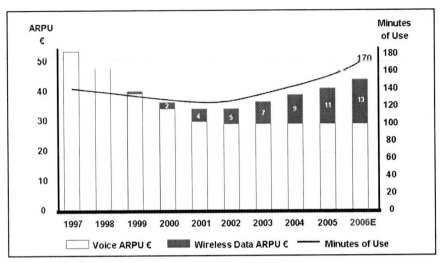

Figure 1: *Arpu*

Today we live in a 3G-technology era (Third generation mobile networks - explained below) with pre 2000 applications. The technology is here. The bandwidth is here. But, the Mobile Data Industry has only achieved only a fraction of its true potential. The missing piece of the puzzle is the 'user oriented applications'. The industry is still hampered by fragmentation: both in terms of technology and the value chain. In contrast to the fragmentation seen in the Mobile Data Industry, the World Wide Web is a unifying force which amalgamates ideas, people and technology on the foundation of open standards.

It seems obvious but could the Mobile Data Industry benefit by following the ethos of the Web?

When OpenGardens first outlined the impact of the interplay between the web and the Mobile Data Industry back in 2004, we were met with considerable scepticism. The book was deemed to be radical, idealistic - and in many ways – wishful thinking. Fast forward one year, and suddenly the concepts advocated in OpenGardens are fast becoming mainstream. Critically, we are seeing changes from both directions i.e. the Internet and the Mobile Internet.

Ajit Jaokar and Tony Fish

Take the case of the Mobile Internet

Broadcast content is losing it's appeal: According to Richard Windsor of Nomura research as reported in the Financial Times (London) on Aug 1, 2005 – 'ringtones are no longer ringing in the tills'. In a nutshell, as ringtones evolve to truetones (which are MP3 songs), most of the income from truetones is going to the record labels. In other words, 'mobile' is just a channel for the media players and the companies who set up services to market ringtones are finding that they no longer have a viable business.

Walled gardens are collapsing: According to Justin Pearse as reported in 'The Feature[3]' – mobile network operators Vodafone and Orange in the UK now admit that between 50 to 70% of their revenues are coming 'off portal'. This means the 'walled gardens' strategy is not paying off as handsomely as the mobile network operators would otherwise have hoped.

But, even bigger changes are happening on the Web itself. The dark clouds of the dot com bust have been swept way. Technology is back in vogue and even the venture capitalists are investing. Advertisers, who originally shunned the Web, are now endorsing the NEW Web with their wallets.

According to GroupM[4], by the end of 2007 in the UK, Internet advertising will close the gap on regional newspapers, the number two medium, but will still be well short of television, the biggest outlet in the £12bn-a-year media advertising market. Excluding Internet spending, total media advertising would be in recession with television, national and regional press reporting revenue falls. Search advertising, in which businesses pay to reach customers trawling on websites such as Google, accounted for more than half of web advertising in 2005.

---

3 www.thefeature.com is no longer operational
4 http://www.groupm.com/

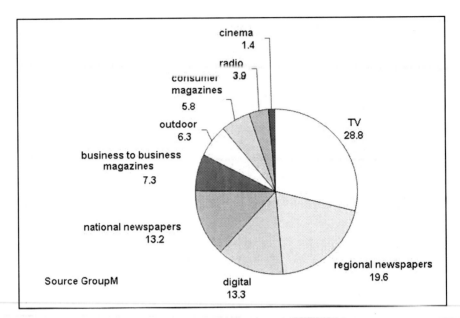

Figure 2: *Advertising revenues*

Many of these changes are being driven by a maturing Web, which differs significantly from the older dot com mania. The catchphrase which encapsulates the 'second coming of the Web' is **Web 2.0**. There is already a media buzz behind Web 2.0, which has arisen because of the wide adoption of broadband and the more sophisticated online services available on the 'semantic' Web. Love it or hate it, it's here to stay!

For us, Web 2.0 provides a common lexicon to explain the concepts of a seamless service experience from the perspective of the web (as opposed to mobile devices). The phrase 'Web 2.0' first originated at an O'Reilly conference in 2004 and then later from a seminal article published by Tim O'Reilly on the Web in September 2005 entitled: 'What Is Web 2.0: Design Patterns and Business Models for the Next Generation of Software[5]'

Web 2.0 comprised of seven principles. We will discuss the seven principles proposed by Tim O'Reilly in greater detail below. However, for now, let's just

---

5 http://www.oreillynet.com/pub/a/oreilly/tim/news/2005/09/30/what-is-web-20.html

Ajit Jaokar and Tony Fish

list these principles. They are:

1) The Web as a Platform

2) Harnessing Collective Intelligence

3) Data is the Next 'Intel Inside'

4) End of the Software Release Cycle

5) Lightweight Programming Models

6) Software above the Level of a Single Device

7) Rich User Experiences

We studied these principles in an attempt to understand more about Web 2.0. At that time, not much was written about the sixth principle i.e. 'Software above the Level of a Single Device'. In an attempt to fulfil this gap, we wrote an article (the first in the series of three articles) on the OpenGardens blog www. opengardensblog.futuretext.com on Dec 25 2005 discussing 'Mobile Web 2.0' i.e. the extension of principles of Web 2.0 to mobility and digital convergence. This article was subsequently republished by Sys-Con media[6].

This series of three articles led to many online and offline interactions. The feedback we received challenged us to think deeply about the issues surrounding Mobile Web 2.0. Our work has been subsequently published at other places including the O'Reilly radar (Web 2.0 – A unified view – April 2006[7]). In retrospect, it was not very difficult for us to write about Mobile Web 2.0, since the extension of the Web on to mobile devices was always on our minds when we wrote OpenGardens.

---

6 http://www.sys-con.com/
7 http://radar.oreilly.com/archives/2006/04/principles_of_web_20_make_more.html

---

In a nutshell, Mobile Web 2.0 is focused on the user as the creator and consumer of content 'at the point of inspiration' and the mobile device as the means to harness collective intelligence

Mobile Web 2.0 is the embodiment of 'Principle six' (Software above the Level of a Single Device) from the list of seven principles for Web 2.0. To truly understand Mobile Web 2.0, you have to study interaction between three distinct industries, all of whom are themselves in the throes of turbulent transformations. The three industries are: Mobile (Telecoms), Media and the Web. The movement of the earth's tectonic plates provides an analogy here. The earth's surface is broken into seven large and many small moving plates. These plates move relative to one another at an average of a few inches a year. The boundaries between these plates is not a pleasant place to be! At the boundaries, we see earthquakes and volcanoes. Sometimes, whole islands can disappear and new islands or mountain ranges can suddenly appear. An apt analogy to the disappearance of old fiefdoms and the creation of new!

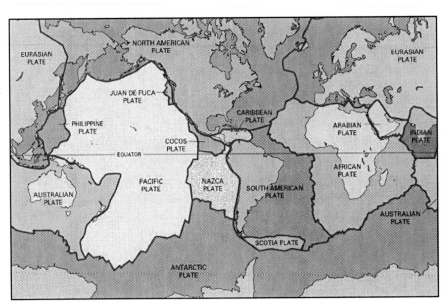

Figure 3: *Tectonic plates*
Source: *http://geology.er.usgs.gov/eastern/plates.html*

                                                        Ajit Jaokar and Tony Fish

The common theme that powers the transformation (tectonic shift) between these three industries is a shift in the creation consumption balance. The first tremors from this tectonic shift were caused by the rise of the Web itself. Web 2.0 is a formal acknowledgement of that shift. Following on from Web 2.0, Mobile Web 2.0 leads to a further dramatic acceleration in the realignment of the creation consumption balance. Mobile Web 2.0 has the capacity to dramatically change the landscape.. as we shall see below.

At this stage, it is worth noting that the focus of our book and our views of Mobile Web 2.0 are driven by the consumer. The reason for this is: whilst we believe that business applications are valuable, viable, growing and extremely important; the market uptake for Web 2.0 in the business domain is slow. In contrast, the consumer market is the early adopter and innovator. This is contrary to traditional new product innovation which has traditionally been led by corporate adoption.

The significance of Mobile Web 2.0

Today, the Mobile Data Industry is broadcast content driven. It has been built from the premise that if it worked on the "silver screen" it will work on all screens including small ones. For the billions of personal and individual mobile handsets, this assumption is not true. There is a creation consumption balance in everything we now do. In the old world, media companies owned and controlled creation and we, the consumers, consumed. It was (and in fact, still largely is) a one-way relationship. That's what we mean by 'broadcast content' driven. In this model, the editors are the most powerful controllers in the value chain because they chose the themes (topics) and then promoted what they thought you wanted (but it was not necessarily what you really wanted!).

The Web has fundamentally changed this creation / consumption balance, as it has opened the possibility for all individuals to become creators of their own content and for others to consume it, without an intervening media company or content controller. In other words, we are now able to choose what news is and what is not. This has been seen with the rise of blogging and sites such as MySpace[8] as these personal web publishing tools have allowed the individual

---

8 www.myspace.com

to publish their own version of what is news. However creative moments occur at any point, and only the mobile device is poised to 'capture, at the point of inspiration'. This makes the humble mobile device the main driving force behind the next evolution of the creation / consumption balance.

No doubt, the PC will also continue to be a place to capture creative moments. But, the PC is not in your pocket and thus it is not at the point of inspiration. Your mobile device is. Of course, there will always be a growing market for push media such as gaming, sports results, horoscopes and ringtones and also for non person-to-person communication like IVR (Interactive Voice Response). However, these will remain insignificant compared to the usage from content captured at the point of inspiration. Further, the PC/web will become the primary point of consumption (entertainment) for all those captured moments of inspiration. The usage of the three main 'screens' in our life is undergoing a change:

Mobile Web 2.0 devices will drive the capture of inspiration. The information captured is not just pictures – but rather a range of things like: calendars (the things you need to remember); notes and reminders (viewpoints and ideas that you want to work on later); news (citizen's reporting) etc. These will be then stored in your private web space. Besides the information you explicitly capture, Mobile Web 2.0 applications will implicitly capture data elements like location and will facilitate access to our private information and to shared public information.

PC 2.0 (for the lack of a better phrase ..) will provide powerful tools for editorial, configuration and cache (local storage).

TV 2.0 (to coin yet another phrase) will be the primary point of viewing all these services. But, TV now becomes just a 'large screen' because the configuration is done at the PC or set top box. The same content could be viewed on the PC screen as well - but TV is just the largest screen with the best rich media capabilities (speakers, surround sound etc).

Thus, this holistic approach will drive a converged media experience. Creation will be driven by the mobile device and consumption by the PC/TV. All the three screens are needed to provide a balance. Business applications could also follow this consumer lead market, but currently the market is still consumer led.

Before we start, a brief note: This book is divided into two parts. Part One discusses Web 2.0, Mobile Web 2.0 and the seven principle of Mobile Web 2.0. Part two covers a range of factors affecting Mobile Web 2.0

PART ONE
The Mobile Data
Industry Today

In this section, we discuss the Mobile Data Industry as it stands today. As we have discussed in the introduction, this traditional view of the industry is 'broadcast content driven', but that view is rapidly changing. The Mobile Data Industry attracts a diverse range of players and understanding the people and their motivation is a good place to start our study of the industry.

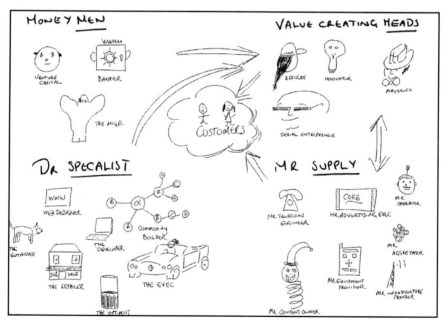

Figure 4: *Players*

You are likely to encounter an interesting and motley bunch of people!

- **The refugee from the dot com industry:** "I lost a (paper) million in the dot com boom but I am going to make my next million here!

- **The maverick:** "Hmm – what have we here?"

- **The innovator:** "I want to be an entrepreneur. I have a single idea / concept which I want to test out"

- **The serial entrepreneur:** The innovator with money

- **The Venture Capitalist:** Who sees it as a new sector to make money but is wary of the stigma of dot com.

- **The Telecoms engineer:** He has been in the Telecoms industry for many years (and yes, it's almost always a 'he'). Knows it well from an Operator/Telecoms point of view. For the first time, he is exposed to the consumer market directly.

- **The advertising/marketing executive:** Aware of brands, Internet, marketing etc. but clueless when it comes to Mobile. Thinks all problems could be solved only if there were the right marketing/branding focus backed with wads of marketing dollars.

- **The media/content owner:** Who owns/manages rights to content. Approaches the industry with a view to monetise that content. Probably already has a similar venture on the Web.

- **The web designer /developer:** Already familiar with the Internet. Seeks to extend his knowledge to the Mobile Internet.

- **The community builder:** "I have built a web based community and I think I can extend that to the Mobile sphere OR I want to create a Mobile community".

- **The enthusiast:** Just interested.

Ajit Jaokar and Tony Fish

- **The retailer:** Got the first taste of success with 'Mobile marketing'. Looking for more.

- **The infrastructure manufacturer:** Telecoms equipment manufacturer. Creates 'Telco grade' applications/ software i.e. software applications meant to be installed at the Mobile operator's premises.

- **The executives:** Who likes to think they understand this space.

- **The development manager:** Who is responsible for third party engagements within a telecoms entity.

- **The Mobile portals/aggregators:** Similar to Internet portals, trying to make themselves the first port of call.

- **The device/handset manufacturer:** Works with a company like Nokia, Sony Ericsson etc.

- **The Telecoms Operator employee:** Not an engineer but in the Telco industry. Works for BT, Vodafone, AT&T etc.

- **A 'point of interest' on the map:** This includes everyone from the art gallery to the corner newsagent who are often at the end of a 'find my nearest' query.

- **The perpetual optimist:** Who hangs out with little cash but with high hopes and finally – the most important person.

- **The subscriber:** Also called the customer. The person, who pays for the service by subscribing to the Mobile operator's network, pays the bill and calls when things don't work. Subscribers are increasingly becoming more than 'consumers'. In some cases, they are becoming producers of content by being active participants in the value chain.

A by-product of all this diversity is the lack of a holistic view – i.e. groups often see the world only from their point of view – sometimes missing the big picture.

More importantly, the customer is viewed as a passive consumer of packaged content. The empowerment and 'Internet awareness' of the customer is the biggest change driving the mobile industry as we shall soon see.

The most important person – the customer

The customer is the one who pays for the service. As with any industry, the goal is to 'serve the customer' – the most important person in the list above. From the customer's perspective, they see two entities – the Device manufacturer (who makes the phones) and the Mobile network operator – who provides the connection to the network. Both are critical players in the industry and are not present on the fixed line Internet. Through the device, the customer accesses a range of services. Primarily, these are voice, since the Mobile phone is still predominantly a voice oriented (rather than a data oriented) device – but increasingly they are data services.

A service the customer accesses is more than the application (software). It could roughly be described as an 'application deployed' i.e. the core application plus support structures such as billing, customer services etc.

Broadly, Mobile data as a service could be classed into a **content based service** or a **contact based service**. The existing (media centric) view of the world is broadcast content based. In this view, the customer is treated as a passive consumer of content. The flow of content is one way i.e. from the media company to the customer via the network. In the contact based service, two or more people communicate with each other via the network. SMS is a good example of a contact based service.

The network

From a customer standpoint, while there are many similarities between web based applications and mobile applications, there are two obvious differences:

- The application is deployed on a mobile device and

Ajit Jaokar and Tony Fish

- The application is accessed 'over the air'.

This means, by definition, at some point – the content must 'fly' i.e. must be transmitted over the air interface to the device where the customer can interact with it. This is achieved through the wireless network. When we are discussing wireless networks, it's important to clarify some terminology.

In Europe, the commonly used phrase for Telecoms data applications is Mobile. In USA, it is wireless or cellular. In this chapter, we use the following terminology:

Wireless: simply implies connection without wires.

Mobility or 'Mobile': Describes a class of applications which permit us to interact and transact seamlessly when the user is on the move 'anywhere, anytime'.

Cellular: refers to the cellular structure of a radio frequency network (explained below)

The Mobile Data industry: The term 'Mobile Data industry' collectively refers to all the terminology and technologies we discuss. The Mobile Data Industry is 'non voice' i.e. refers to only data applications on mobile devices

Hence, we use the term 'Mobile' independent of access technology i.e. 3G, wireless LANs, WiMax, Bluetooth etc. The wireless network itself comprises of the actual physical network which facilitates the air interface. A wireless network can range from a personal area network (for example - a Bluetooth network offering coverage of 10 to 100 meters) to a satellite network which offers global coverage. As an applications developer, we can often treat the lower level functionality of the network as a 'black box'. If needed, such functionality can be accessed through defined APIs in cases where they are publicised (not all functionality of the network will be accessible to applications due to privacy and security reasons).

The four classes of wireless networks are:

- Personal area networks (for example Bluetooth)

- Local area networks (for example Wireless LANs)

- Wide area networks i.e. the mobile operator managed radio frequency networks and

- Satellite networks

While Satellite networks are out of the scope of this chapter, the first three networks can be classified into two broad subclasses.

1) Localised networks and

2) Wide area networks i.e. the Mobile operator managed Radio Frequency network (RF network)

Localised network	Wide area network
Based around an access point/ hotspot	Can be accessed anywhere independent of a hotspot.
Unlicensed band	Part of a licensed spectrum
Ex: WiFi ,Bluetooth	Ex: Cellular networks

Localised networks

Localised networks are created around a hotspot/access point and have a limited range within the proximity of that hotspot/access point. These networks include the WiFi network and the Bluetooth network. Unlike wide area networks (explained below), localised networks operate in the unlicensed spectrum (hence, they are 'free'). In contrast, wide area networks (RF networks) operate in the licensed spectrum (hence, the high cost of 3G licences in Europe). Since WiFi and Bluetooth are two common implementations of localised networks, we shall discuss them in some detail below.

WIFI or Wireless LANs is a term which refers to a set of products that are based on the IEEE 802.11 specifications. The most popular and widely used Wireless LAN standard at the moment is 802.11b, which operates in the 2.4GHz spectrum along with cordless phones, microwave ovens and Bluetooth. WiFi enabled computers or PDA (personal digital assistants) can connect to the Internet when in the proximity of an access point popularly called a 'hotspot'. The WiFi Alliance [9] is the body responsible for promoting WiFi and its association with various wireless technology standards.

The Bluetooth network: Bluetooth is a wireless technology specification, which enables devices such as Mobile phones, computers and PDAs (Personal Digital Assistants) to interconnect with each other using a short-range wireless connection. It is governed by the Bluetooth SIG[10] (Special Interest Group). The operative word is 'short range'. A typical Bluetooth device has a range of about 10 metres. The wireless connection is established using a low-power radio link. Every Bluetooth device has an in-built microchip that seeks other Bluetooth devices in its vicinity. When another device is found, the devices begin to communicate with each other and can exchange information. Thus, a Bluetooth enabled device can be thought of as 'having a halo' seeking to communicate with any device that enters the range of that halo. Bluetooth is 'free', in the sense that it is an extension of the IP network in an unlicensed band via the Bluetooth access point. Although Bluetooth has not lived up to much of its initial hype, the technology is significant since most phone manufacturers have committed to Bluetooth enabled phones.

The Radio Frequency (RF) network

In contrast to localised networks, the RF network is not confined to specific hotspots/access points. The RF network is a 'cellular' service in the sense that the actual network can be viewed as a honeycomb of 'cells'. The basic cellular network has been used for voice transmissions since the 1980s and subsequently for data transmissions since 1990s. The entity that manages the cellular network is called the Mobile Network Operator (also called 'operator'

---

9   http://wi-fi.org/OpenSection/index.asp
10 www.bluetooth.org

or carrier). Examples of mobile network operators include T-Mobile http://www.t-Mobile.com/, Verizon wireless http://www.verizonwireless.com/ and NTT DoCoMo http://www.nttdocomo.com/. There are two ways we can categorise cellular data transmission technologies:

a) By understanding their historical evolution and

b) By understanding the cellular data transmission techniques.

The historical evolution of networks (i.e. (a)) is more familiar to the general public through terms like '3G' etc.

Historical evolution of data transmission techniques

Cellular systems can be viewed as the 'generations' of systems as they evolved over time (2G, 2.5G, 3G etc.). The main differentiation across the generations is support for greater bandwidth. Obviously, as you go towards 3G and beyond – the bandwidth increases and the applications supported also become richer. The first generation (1G) systems were analogue systems. From 2G (second generation) onwards, the cellular systems have been digital. We will not discuss analogue systems in this book and hence we start our discussion from 2G systems onwards.

2G systems

GSM (Global System for Mobile) is the most popular 2G system. GSM originated in Europe and is the dominant Mobile system across the world. In some form, it is present in all continents including North America. GSM (based on TDMA) is a digital system with a relatively long history (the study group was founded in 1982) and is governed by the GSM Association (http://www.gsmworld.com/index.shtml). The GSM Association provides functional and interface specifications for functional entities in the system but not the actual implementation. Besides GSM, other examples of 2G systems are cdmaOne (mainly in the USA), PDC (Personal data cellular) in Japan.

2G technologies are typically capable of supporting up to 14.4Kbps data. 2G systems are characterised by being 'circuit switched' (i.e. a circuit is first established between the sender and the receiver before sending the information and is maintained for the duration of the session). The next evolutionary step (2.5G described below) is characterised by being 'packet switched' (the data is broken into packets and no connection is maintained for the duration of the communication). Note that both circuit switched and packet switched networks are digital in contrast to previous analogue systems.

2.5G systems

2.5G networks are an intermediate step undertaken by most Mobile operators in their evolution from 2G to 3G. The main functional leap between 2G and 2.5G networks is the adoption of packet switched technologies (in 2.5G networks) as opposed to circuit switched technologies (in 2G networks). 2.5G networks are capable of theoretically supporting bandwidth up to 144kbps but typically support 64Kbps. GPRS (General Packet Radio Service) in Europe and CDMA2000 1X in North America are examples of 2.5G networks. Applications such as sending still images are possible over 2.5G networks.

3G systems

Most people have heard about 3G and have an opinion about it! If nothing else, they know about 3G in terms of the high prices paid by Mobile operators for 3G licenses. However, from a Mobile operator perspective, there is a clear business case for investing in 3G because existing 2G networks are congested 2.5G solutions are a 'half way house' and will not cope with the increasing demand (i.e. both the number of consumers and the richer application types). From an application development perspective, 3G technologies are differentiated from 2.5G technologies by a greater bandwidth (theoretically 2Mbps but typically 384Kbs). Possible 3G applications include video streaming.

From an end user point of view, the move from 2.5G networks to 3G networks is more evolutionary than revolutionary except in the case of devices. 3G devices are significantly more complex because of the need to support complex data types like video, provide more storage, and support multiple modes. UMTS (Universal Mobile Telecommunications System) in Europe and CDMA2000 in the North America are examples of 3G systems. Note that 3G systems are all based on CDMA technologies.

Type	System Characteristics
2G	Capable of supporting up to 14.4Kbps data. 'circuit switched' Example GSM
2.5G	Packet switched Theoretically supporting bandwidth up to 144Kbps but typically support 64Kbps. Example: GPRS (General Packet Radio Service) in Europe and CDMA2000 1X in North America
3G	Packet switched Theoretically capable of supporting 2Mbps but typically support 384kbs

Cellular data transmission techniques

In addition to historical evolution of networks, cellular systems can also be classified based on the data transmission techniques. There are two main techniques for cellular data transmission – TDMA (Time division multiple access) and CDMA (Code Division Multiple Access). The discussion of these technologies is out of the scope of this book.

The classic value chain

All the above players could be depicted in the traditional value chain for the Mobile Data Industry shown as below:

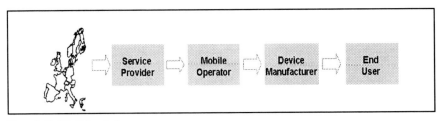

Figure 5: *Mobile Value chain*

The players in the value chain are as depicted above.

Content owners: Content owners include players like Broadcasters (TV, radio etc.), News agencies, Publishers, Entertainment companies (movies, music, entertainment), Rights Owner Companies (music rights, sports rights, general showbiz agents etc.). Unlike the Internet, the general perception of the Mobile Internet is that content is not free. Content owners own the rights and /or represent the copyright owners. Content owners believe that 'Content is King'

Service providers /aggregators: Service providers act as an aggregator in the market. A service provider could be a mobile operator or an independent portal. They are concerned with billing, customer support etc. They could be customer facing. They believe in high volumes ('Pile 'em high – sell em cheap') and could work on a revenue share model but would like upfront payments where they can.

Mobile network operators: Mobile network operators actually manage the physical network. And in some cases, also fulfil the aggregator role. The mobile operators have a direct relationship with the customers and influence the whole value chain.

Device manufacturers: Devices/Device manufacturers: are often the first physical point of interaction with the customer. Examples of device manufacturers include Nokia, Sony Ericsson, Samsung etc. They can be strong brands in their own right. Device manufacturers are trying to own a larger share of the value chain by becoming portals. For example - Club Nokia www.nokia.co.uk/clubnokia from Nokia.

End Users: The end user is the actual consumer of content and pays for the content. The end user defines the market demand.

Content Is King?

Introduction

As we have seen in the previous chapter, the classical value chain for the Mobile Data Industry points to a dominance of the media industry. It indicates a symbiotic relationship between the media industry and the Mobile Data Industry based on the notion of 'Content is King' because the Mobile Data Industry has become another distribution channel for the media to deploy their content.

But, is Content King?

We will discuss below that content is NOT king. In fact, this argument (content is NOT king) has been comprehensively proven as early as January 2001. In a seminal paper dating Jan 2001, the mathematician Andrew Odlyzko (then at AT&T Bell labs, but now at the University of Minnesota) introduced us to the concept that content is NOT king[11] . The date of this paper is worth noting (Jan 3, 2001), which is at the height of the dot com boom when contrary viewpoints prevailed. He argues successfully that contact and not content is the real differentiator.

We are interested in exploring the idea that contact (and not content) is king because person to person communication (i.e. contact) is a key driver for Mobile Web 2.0.

As we alluded to in the introduction with the example of the changing emphasis of the three screens (TV, PC and mobile), it is not possible to look at one sector

---

11 http://www.dtc.umn.edu/~odlyzko/

(such as Mobile) in isolation. In fact, with increased digitisation, the once mythical notion of digital convergence is a fast becoming a reality (as we shall see in subsequent chapters of this book). Hence, to understand the Mobile Data Industry as it currently stands; we need to understand the thinking behind the major players in the media industry and the current symbiotic relationship between the media industry and the Mobile Data Industry. Finally, we need to understand the seismic changes within the media / content industry which are driving both Web 2.0 and Mobile Web 2.0. We can view this change as appearing in waves, much like the seismic waves after an earthquake.

> Note: In this chapter, when we refer to 'Content', we are referring to 'broadcast content' unless otherwise stated.

The history of 'Content is King'

Predictably, the idea that Content is King is publicised widely by Hollywood and other providers of packaged content. For those who claim that 'Content is King', 'content' means movies, sports events, music and other pre- packaged content. However, Andrew Odlyzko argues that the annual revenue from connectivity (contact) are far greater than content when he says:

> *The annual movie theater ticket sales in the U.S. are well under $10 billion. The telephone industry collects that much money every two weeks! Those "commodity pipelines" attract much more spending than the glamorous "content."*

While the idea of Content is King was widely publicised by the movie and recording industry, it originated much before that. Again, Andrew Odlyzko points to a much earlier incident with the US government's subsidisation of the newspaper industry in the early 19th century. The then prevailing logic was: newspapers were needed to keep the nation informed, engaged and educated. Hence, newspaper distribution was subsidised by the profit from letters (letters being a person to person contact mechanism). However, according to the paper, there were reasonable arguments that the preoccupation with newspapers harmed the social and commercial development of the country by stifling circulation of the informal, non-content information that people cared about.

In other words, the high prices for stamps (driving person to person communication) were caused by the subsidy to the newspaper industry (packaged content). In the 1840s, when Congress did drop prices of stamps, usage patterns changed dramatically. Another way of looking at this is - *the Post Office would have thrived on letters alone, but would have gone bankrupt instantly had it been forced to survive on newspaper deliveries.* Thus, while the policy makers thought 'Content was King', the customers certainly did not think so, because they were unwilling to pay for it!

The economic factors driving the content industry

Traditionally, the media/content industry has been very complex with geographical variations in 200+ countries. Its business model has been similar to companies selling products, with 'content' replacing 'product'.

Hence, the content industry is currently based on specific products (for example - a movie), complex distribution channels based on physically shipping goods (for example - movie theatres which received film to play in theatres) and a marketing budget to support the product.

The value chain could be depicted as in figure 6. Note that, this value chain mirrors practically any 'physical product'.

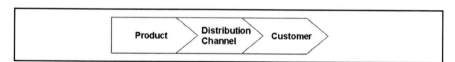

Figure 6: *Product to customer*

The product based model, when applied to content, produces some unique effects. For instance, film studios lose money on most films. New productions are risky. According to the MPAA (Motion Picture Association of America)[12], the average cost to make and market an MPAA film was $96.2 million in 2005. This includes $36.2 million in marketing costs. Thus, the marketing costs were about 40% of the movie's total costs. The high overhead, upfront costs of

---

12 www.mpaa.org

production and high distribution costs stifles innovation. Thus, when MGM[13] was bought by private equity firms, their remit was to focus on the market exploitation of the archive (older films) and this limited new productions.

The result is, Hollywood has become a 'winner takes all' industry. (The same logic applies to record labels as well). This means, the current 'product based' value chain supports a few big hits and expects many losses. In other words, the Hollywood system as it stands cannot currently monetise the 'long tail', as we shall see below. While Hollywood and the recording industry make all the waves in terms of press coverage, the overall media industry itself is much larger. The components of the global /infotainment market ranked by size include:

- TV distribution

- Newspapers

- Broadcast / cable networks

- Filmed Entertainment

- Internet Access

- Magazines

- Radio

- Video games

- Consumer Books

- Recorded music

---

13 www.mgm.com

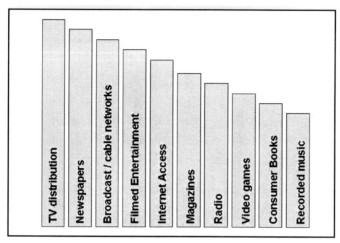

Figure 7: *Components of the global infotainment industry*

Each component has largely existed in its own commercial silo, following its own product-distribution-customer model. Thus, the media industry as it stands today, is burdened by overheads (for example: product managers for each segment etc.) which originate from the concept of selling physical products. Further inefficiencies exist in the value chain when we look at other components like regulation, copyright, technology, payment mechanisms, standards, geography etc.

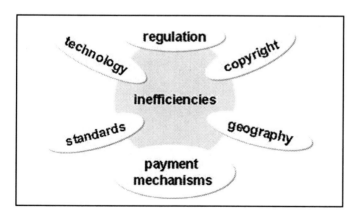

Figure 8: *Inefficiencies in the media industry*

Thus, the media industry needs significant financing and looks for ways to monetise its content to fund its rather arcane model. The high investment in the industry is recouped from a number of sources such as:

Brands: Brands engender trust and hence command a premium. Brands also protect against the downside; hence, the premium paid to top actors (who are brands in their own right). This model is self fulfilling because the industry has a very high cost base and inbuilt inefficiencies. It needs assurance to recoup its investment. This leads to high premiums for a few players.

Copyrights and patents are another source of value and a potential method to recover the investment in the industry.

Additional revenues are obtained from merchandising and carefully controlled releases for different media domains (media types, geography etc). Movies windows dictate the business models since revenue is maximised through managed distribution channels. The initial window has been domestic exhibition (cinemas), followed by DVD release, international release, cable movies etc.

Other sources of income include sponsorship, advertising, government aid and product placement etc.

To summarise the above discussion:

> High investment costs and an inefficient distribution mechanism support a small number of hits. Controls and barriers are put in place to recover this high investment cost.

We are now going to consider the three seismic shifts.

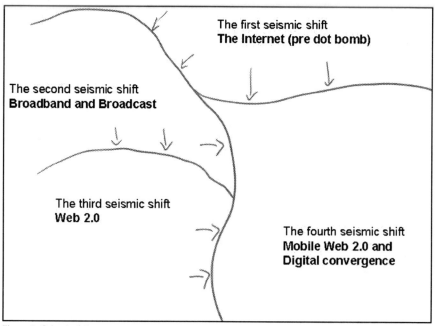

Figure 9: *Seismic shifts*

The first seismic shift - The Internet (pre dot bomb)

The dominance of packaged content continued for a long time – until the Internet came along. The Internet changed everything and there was no going back!

The first wave of the Internet revolution spanned 1996 to 2001 i.e. before the dot com crash. From the media industry standpoint, this was characterised by a dramatically more efficient distribution model. The new model was low cost and removed all the inefficiencies of the existing distribution model (bookstores, product managers etc). However, note that the general public took a lot longer to get used to this change and they were still using dial up connections to the Internet in a majority of the cases. **Thus, the darling of that era was Amazon (www.amazon.com) which capitalised on an efficient distribution model but was still selling physical media (books), given that selling electronic products was limited by bandwidth.**

This era also introduced us to the first glimpse of a business model which could work in the future i.e. advertising sponsored content. Although the model failed in the first phase of the Internet, the advertising model is making a successful comeback in subsequent phases of the industry. This is because faster broadband means consumers consume online services faster and in greater numbers thus making advertising more compelling.

The second seismic shift – Broadband and Broadcast
Broadcast vs. Broadband

Despite the doomsday pundits, the Internet itself survived the dot bomb. In fact, there is an argument that it became much more useful after the speculators who had concentrated on traditional models exited the market. The two biggest characteristics of the post dot com era were: 'The rise of broadband' and 'A return to communications'.

Broadband Internet access (or broadband for short) is a high data-transmission rate Internet connection. Typical broadband rates are 256 kilobits per second which is about four times those of the telephone modem (typically 56Kbps). Modern consumer broadband implementations are up to 30 Mbit/s. Meanwhile, broadcast has been a major component of the Infotainment / content market. As broadband Internet becomes more widespread, broadcast and broadband are increasingly in collision. Broadcast and broadband achieve the same function i.e. serving content to the customer. But, they do it in different ways. Broadband functions without the overhead and regulation of broadcast. Broadband also offers customers a choice, which broadcast does not because broadcast is predominantly 'push' media.

Regulation is a major component of the broadcast / content industries.

Generically broadcast was dominated by terrestrial broadcasting operators usually public service and government owned until commercial channels were introduced. Barriers to entry were high guaranteeing oligopolistic positions. However, now spectrum is being reorganised, coupled with increasing digitisation which creates interesting new dynamics for content. New competition in transmission media is emerging for example through Digital

MultiMedia Broadcasting (**DMB**) in Korea, Internet Protocol TV (IPTV) and **DVB-H** (Digital video broadcasting – handheld).

But, the biggest challenge to broadcast is broadband itself, performing the same function as broadcast, but potentially more effectively and giving the customer a lot more choice. The consumer sees broadband as a 'fast' data connection. But, Broadband does more than provide a fast connection. It connects cameras, moving image servers, producers and users allowing the delivery and sharing of content at lower prices. Now images, films and music may be delivered to / accessed by consumers without the physical overhead.

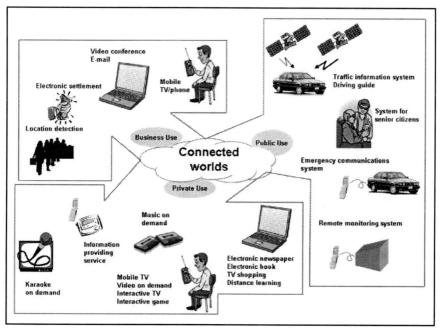

Figure 10: *The potential of broadband*

This allows content to be shared in new ways. No longer is content the sole province of the publisher and broadcaster. The new distribution channels have meant that many producers can access the same consumer in more ways. More importantly, consumers can access and use each others' work in new ways by-passing commercial interests. In response broadcasters have developed their own websites to repurpose content ranging from news and weather to moving image and sound from the TV/Radio medium to Web based.

For example www.backstage.bbc.co.uk, is the BBC's developer network to encourage innovation and support new talent. Content feeds are available for people to build user generated content on a non-commercial basis and Channel4's site - http://www.channel4.com/fourdocs is a place to upload or download four minute documentaries created by consumers. Anyone with a story to tell or an opinion to voice can submit their film to FourDocs.

Similarly print media is having to deal with a new wave of user generated content. For example: http://www.ohmynews.com is a Korean (includes English International version) online newspaper, where the majority of the content is supplied by the general public. Its 10 million users make it the sixth most influential media in Korea and the most important media for younger people. It is having a dramatic effect on the nature of journalism, shifting news from print and television to online and challenging the way that news is gathered and reported – its motto is 'every citizen is a reporter'; of whom there are 35,000.

The shift to broadband is just beginning and it still has a long way to go especially in the next few years leading up to 2010. Generally, the total switchover from analogue to digital television is set for 2010+ with most countries moving to High Definition Digital TV in that time frame. The World Cup in mid 2006, the Beijing Olympics in 2008 to be followed by analogue switch off (2010+) leading up to the Olympics in the UK in 2012, will all act as a stimulus on content.

There is intense debate about how the emerging IPTV, VOD (Video On Demand), Internet Content on Demand (ICoD), and Mobile TV services will evolve and the impact that they will have on broadcast services. A variety of different models for new personalised, interactive, on-demand TV-like services are being developed by fixed and mobile operators, ISPs and broadcasters, based on a range of different technologies, platforms and devices. However, widespread service deployment has been delayed due to regulatory uncertainty. Over the next few years, that regulatory uncertainty will clear up leading to a boost for broadband adoption.

Drivers and economics

We therefore have the following drivers to broadband:

- Increasing competition in the broadcast industry from a wider array of digital broadcasters managed by regulators.

- Analogue to digital transition plans of governments worldwide often stimulated by major sporting events such as the World cup and the Olympics.

- Adoption of broadband by advertisers as a medium to reach their customers.

- Government regulations which create the operating environment for content. They may be motivated by consumer issues, cultural issues or international competitiveness drivers.

- Governmental policies to decentralise the content industries from capitals to the regions.

- Broadband pricing impacts on consumer market uptake, content delivery charges and business models.

- Broadband availability, capacity and price impacts on target customer markets e.g. urban or rural, high speed or low speed.

- Content producer's window management i.e. how attractive is broadband as a window to reach customers.

- Rights management regimes. How content procurers and deliverers manage their rights management regimes dictate how the content markets evolve.

- Spectrum management dictates the economics relating to the delivery of content over the spectrum. This may skew economics benefits towards one group in the value chain as opposed to the government or the consumers.

Implications on the big picture

With broadband, both customers and content producers have a choice.

Content producers have new and more efficient distribution channels. Customers have more choice and more variety of channels (such as through mashups). Content need not be tied to the device (such as TV, PC etc). Most importantly, customers need not mindlessly consume products heavily promoted by the large companies. They have much more choice and control over when they can consume the content and how.

The third seismic shift – Web 2.0

While broadband is a facilitating technical medium, Web 2.0 is the shift in customer behaviour. It is based on 'harnessing collective intelligence' and could be viewed as the 'Intelligent Web'. We will discuss Web 2.0 in the next section, however – there are two facets of Web 2.0 that we will introduce in this section since they pertain to content management: the 'long tail' and the 'community as the editor'.

The long tail

In most situations, 80% of the revenue comes from 20% of the products/ services (depicted in red below). Thus, the remaining 80% of the products have low demand and low sales. These constitute the 'long tail'. The principle of harnessing the 'long tail' argues that collectively in a digital environment where there is no limit on storage archives, these low volume/low sales products can make up market share that equals or exceeds the few bestsellers – provided the distribution channel is large enough and the per unit production cost is low. The long tail is depicted shaded as below.

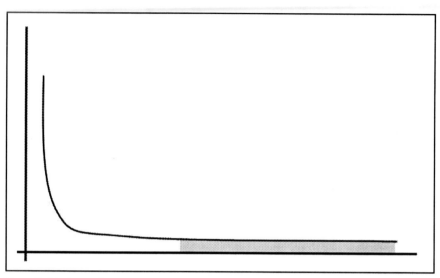

Figure 11: *The Long tail*

The principle of the Long tail manifests itself in many ways with Web 2.0; but for now, let's consider its impact on the movie business. We have seen previously that, according to the MPAA (Motion Picture Association of America)[14], the average cost to make and market an MPAA film was $96.2 million in 2005. This includes $36.2 million in marketing costs. Thus, the marketing costs were about 40% of the movie's total costs.

However, at the other end of the spectrum are small, independent films that have become hugely successful. These include: El Mariachi ($7000), the Texas Chainsaw Massacre ($140,000), Mad Max I ($400,000) and the Blair Witch Project ($35,000)[15]. The commercial acceptance of these small budget films proves that it is not always necessary to spend a lot of money on big budget films. These films, and many others, are examples of the long tail. Although many films coming under the banner of the long tail will not be as hugely successful as the films listed above, collectively, they do offer sufficient revenue to create a business.

---

14 www.mpaa.org
15 http://www.empireonline.com/features/50greatestindependent/46_45.asp

The community as the editor

One of the criticisms of the long tail theory is its lack of selectivity i.e. the big media companies often argue that customers are faced with too much choice and will not have the time to select the 'best' content. Hence, the need for an editor/selector/gatekeeper (translation: themselves!). Many Web 2.0 sites are increasingly disproving this fanciful notion because the community is becoming the editor. Two instances that show that the community can be an excellent editor are:

a) **Digg (www.digg.com)** According to the Digg website, Digg is a technology news website that combines social bookmarking, blogging, RSS, and non-hierarchical editorial control. With Digg, users submit stories for review, but rather than allow an editor to decide which stories go on the homepage, the users do. The more 'Diggs' a story gets, the more it comes up on the front page. Does Digg work? On Nov 17, 2005 - wired magazine called Digg the 'Slashdot killer'. Slashdot has been established for seven years and is regarded as an authority in itself. Digg's 80,000 userbase is doubling every three months.

   Source: http://www.wired.com/news/technology/0,1282,69568,00.html

b) **YouTube (www.youtube.com)** YouTube is a consumer media site for people to watch and share original videos worldwide through the web. Users can easily upload, tag, share and rate video clips through youtube. com. YouTube serves up over 25 million videos daily and counts Sequoia Capital as one of its investors.

We believe that this is a part of a larger trend where long tail and community editing will complement each other.

The fourth seismic shift - Mobile Web 2.0

Following on from Web 2.0, Mobile Web 2.0 will amplify the effects of Web 2.0 by capturing ideas at the point of inspiration. This book is all about Mobile Web 2.0; hence, this section is brief since we cover Mobile Web 2.0 in extensive detail later. Due to the initial success of ringtones on mobile devices, it was assumed that mobile devices were just another avenue to sell pre packaged content. Indeed, to an extent, that is the case. Thus, the current focus of the media industry leans towards pushing content on mobile devices.

Pushing content on devices through the mobile network has limitations as the content grows richer. Sending a large movie over a limited bandwidth connection would take many hours. Mobile devices are capable of far more than carrying ringtones and other entertainment type content. Specifically, they are all about capturing information. It is this facet of mobile devices which makes them very interesting from a Web 2.0 perspective.

As we shall see later, mobile devices are crucial to Web 2.0 for two reasons: firstly because they are capable of capturing ideas at the point of inspiration and secondly because they are likely to be 'most often used' to capture content. Mashups could be used to drive digital convergence. We cover these ideas in greater detail in subsequent sections.

Winds of Change and a word of caution

We have seen a historical perspective of the content industry and discussed the trends driving its evolution. In this final section, we wrap up this discussion with some trends affecting the industry's future. Before we do so however, a word of caution about attitudes within the Mobile Data Industry.

A word of caution

The Telecoms environment and its slow moving, complex decision-making processes perplex most developers. Hence, a word of caution is needed. When we first discussed the idea of OpenGardens with a well-known Telecoms industry veteran, he said that the book should come with a 'Health warning'. Keep away! Too many have gone down this path with visions of gold rush 'la Klondike' but have died a snowy death. Here is a brutally frank paraphrased view point, coming from a Telecoms industry insider:

"Telecoms has always been a profitable business for the Operator and this extends to the Mobile network. In its early stage, and even today, Mobile is still more about voice than about data.

In most countries, Telecoms/Mobile have been a monopoly – with simple services, which did not need to be marketed actively.

In spite of the setbacks from the dot-com mania and the 'momentary lapse of reason' with 3G licenses, voice is still popular – both on the landline and the Mobile environment and will continue to be so.

In the dot com world, there were calls to 'subsidize Telecoms' for the 'greater good'. No industry can afford to do that and in retrospect, the proponents of 'free' are assigned to the digital graveyard of the dot-com industry. With the rise of the Internet, the fixed line Telecoms operators were relegated to being 'data pipes'. (In the sense that the big stars of the Internet were Yahoo, Google, Amazon etc. who sold services over a network operated by the fixed line Telecoms operator. These businesses were far more profitable but the fixed line Telecoms operator ended up being the 'data pipe' and seeing little revenue.)

When Mobile data came along, initially, there were hopes of 'we missed the Internet, but we will NOT miss this one!' There was, and still is, a tendency to 'go it alone' without partners. It is debatable as to how much value partners add - unless they happen to be well known brands like Warner and Disney. Most ideas are not new and most concepts presented to Mobile operators are merely at the concept/planning stage."

Could there be a new business model for content?

Despite outdated viewpoints like the above, change is happening at a lot faster rate than the telecoms and media industry would like. Both industries are on the defensive with the resurgent web and more mature business models. Surprisingly, things which were written off in the first stage of the Web (dot-com), are making a comeback. Take for instance the advertising model. Once dismissed as unviable, the advertising model could well subsidise an independent film. We have seen from examples previously that a large proportion of the movie budget goes to advertising, distribution and promotion.

A budget of less that $5 million dollars for a movie could well be funded by advertising. In other words, if production costs are low, distribution costs are almost zero due to the Internet, then many more movies could be produced and these may be funded by advertising. Thus, the media industry may see much more turbulence in the near future. The Web 2.0 world and Mobile Web 2.0 world will see a growing, but flat market, for content; many more modest hits than a few massive hits and many more losses.

Digital Rights Management

Every piece of content has rights attributed to it by aspects of copyright and other laws which vary by country law. Some may be at zero prices; some may be expired or have public interest rights providing more freedom for use. For most people, it is a confusing arena. Digital Rights Management (DRM) is the management of these rights in a digital domain. It is often confused with content management.

For mobile DRM and Mobile Web 2.0, DRM has a unique significance due to its impact on the network effect - which we shall explain below.

The network effect, also known as Metcalfe's law, (for the purposes of this discussion, we can use these two terms interchangeably) was the main driver behind the dot-com model. While Metcalfe's law has inevitably been tarred by the failure of the dot-com bubble, the benefit of hindsight shows us that the network effect is indeed a valid phenomenon (albeit misapplied in the dot-com context)

According to Wikipedia
> *The network effect causes a good or service to have a value to a potential customer dependent on the number of customers already owning that good or using that service. Metcalfe's law states that the total value of a good or service that possesses a network effect is roughly proportional to the square of the number of customers already owning that good or using that service.*

S. J. Liebowitz and Stephen E. Margolis elaborate more in their paper Network Externalities (Effects) by giving an example of a fax machine. [16]

In effect, the value availed from a fax machine has two components; firstly the 'intrinsic value component' of a single fax machine but also 'the network effect component' i.e. value gained by other users using fax machines (i.e. the ability to communicate/the network effect). The concept of network effect is thus tied to the concept of 'critical mass' i.e. at a certain point called as the critical mass, the benefits gained from the 'network effect component' kick in and lead to huge competitive advantage and increased revenue. In short, a successful (viral) application.

Until that happens, the full potential of the technology/service cannot be availed and the user base is confined only to early adopters. The dot com boom (and bust) adopted the above principle (which is valid) and added to it the concept of 'unprofitable enterprises' (which is invalid). However, the benefits of network effect are not in doubt. In fact, with hindsight, we can see that the network effect works (think success of Hotmail).

How does this phenomenon play out in the Mobile Data Industry?
Clearly, if we find a set of applications/technologies that have a tendency to benefit from the network effect, we could all benefit. But sadly, at the moment - we see a number of 'failures' rather than successes. The key 'success' in terms of the benefits of network effect is the humble old SMS. In fact, the success of SMS (and also more broadly the success of GSM - which is a collaborative venture) shows that the industry indeed can 'get it right'.

How do we then explain the many failures (i.e. why do we see so few 'network effect' applications?

Taking a step back, let's see what factors contribute to the proliferation of the 'network effect component'? The usual factors prevail - the service must be cheap, easy to use, the customers must want it.. and so on. **But, over and above all this; the service must be 'interoperable' or 'capable of being viral'.** In other words, it must be capable of spreading. If we introduce barriers

---

16   http://wwwpub.utdallas.edu/~liebowit/palgrave/network.html

to the spread of content - we kill it even before we even start. Forget about network effects!

Take MMS (Multimedia Messaging Service), clearly this technology/service could benefit from the 'network effect' component. In simplest terms, it should be easy to 'share' content with other users. This means more people will take up MMS if they can firstly create their own content and secondly share that content (cheaply and easily). Creation of their own content is possible via camera phones; sharing it cheaply and easily is far from possible due to a clunky interface and technical complexities! So, who uses it (if at all)? Inevitably, it's early adopters who have found a better way (sharing images by Bluetooth or email) but it's not the mainstream user.

If the benefits of the network effect are apparent from the Internet, why does the Mobile Data Industry create so many barriers?

Could it be because they are driven by the media/content industry with its protectionist/broadcast/DRM mindset?

Indeed, we believe that the requirements of the media/content industry are in contradiction to the 'network effect' application. In the former (media industry), you must restrict the free flow of content in order to make it more valuable. In the latter (applications benefiting from the network effect), you must actively encourage the free flow of 'user created content'. Note this is not an argument for the 'Napster mindset'. We are not advocating swapping of 'Hollywood's created content but rather seek to encourage the free flow of 'user created content'. At the moment, the restrictions put in place serve the media industry at the expense of the 'real customers' i.e. those who want to communicate. And this cripples the Mobile Data Industry.

The Mobile Data Industry must realise that it's far bigger than the 'song and dance' crew. It is a major communications platform and must not let outdated business practises hamper it's true potential. At the moment, the industry is facing the media barons with its back to the customer! It is driven by the 'broadcast content' model. The 'Lord of the Rings' ringtone, 'Harry Potter' images and so on. All are driven by the music industry/Hollywood. The chase for media dollars serves a short-term goal. Operators must realise that they are the masters and not the slaves to the media industry.

When we speak to operators, they often naively talk of 'on portal/off portal' and insist that there are no walled gardens in the industry today. But that's very limited thinking. Sadly, walled gardens, DRM etc exist to protect broadcast content – often at the expense of true communication! That's the real pity of DRM within the Mobile Data Industry.

The winds of change

But things are rapidly changing in the content industry..

- Oscar award winning director Steven Soderberg launched his new movie 'Bubble' simultaneously in theatres, DVD and cable. This cuts down release expenses for the movie producers.

- The first album from indie band 'Arctic Monkeys' has become the fastest-selling debut album in UK chart history selling more than 360,000 copies at launch. The band built up their fan base on the Internet, after they put their demo MP3s/CDs on the web for other people to hear.

In retrospect, when we talk of the changes due to mobile devices, the winds of impact were felt much earlier, although few recognised it because we were all still feeling the shock waves of the dot-bomb. Way back in August 2003, the Independent newspaper in the UK reported that movie flops like 'The Hulk', 'Charlie's Angels' and others were attributed to teenagers texting friend from **inside** the theatre that the movie was no good! All the careful planning, hyped up product launches and media spin collapsed within the time it takes for a teen to compose a text message. The same fate befell Arnold Schwarzenegger's movie Terminator 3. In the words of Arnie himself – it was *'Hasta La Vista baby!'* [17]. The telecoms execs fearing their industry being relegated to 'pipes' need not have worried.. it appears that they may not be piping things of much value to the customer anyway!

---

17 http://blogcritics.org/archives/2003/08/31/124903.php

Web 2.0 and Mobile Web 2.0

WEB 2.0: Of elephants and wise men

Before we discuss mobile Web 2.0, let us first understand what is Web 2.0 itself. Because of our viewpoint on Mobile Web 2.0, we are asked the question: What is Web 2.0? . It is often a genuine question, since Web 2.0 is often perceived as a marketing tool that has caused considerable confusion in the industry. Further, Web 2.0 is a 'soft concept': it's not a standard, or a formula or a definition, which from our stance would have been a lot easier to explain. Because Web 2.0 is in the news, everyone has an opinion about it. At times, it seems that, like religion and politics, it is prudent to avoid discussions about Web 2.0 altogether and easier to be an agnostic or sceptic rather than a believer!

In our view, the best way to understand Web 2.0 is to study it from it's first principles, i.e. the seven principles of Web 2.0 as outlined by Tim O'Reilly in his seminal document http://www.oreillynet.com/pub/a/oreilly/tim/news/2005/09/30/what-is-web-20.html?page=1. It is important to **collectively apply** these seven principles i.e. to view them as a whole and understand the value added by each observation/principle.

Many people want to take a one dimensional view when it comes to Web 2.0 i.e. like the story of the proverbial blind men and the elephant, they look at only one facet of Web 2.0 and insist that it's the whole. If the blind men looked at the elephant in isolation, they would not perceive it correctly (for example – the elephant's trunk could be thought of as a snake and so on). It is only by considering the totality of all the aspects, can they reach the correct conclusion.

Figure 12: *The blind men and the elephant* © *Jason Hunt 18*

Here are a couple of examples: A person who we met a recently insisted that Web 2.0 is 'community'. Indeed, Web 2.0 has some community features: but no it's not ONLY about community. There is a lot more to Web 2.0 than a traditional community. In another instance, a person believed that their software is in 'perpetual beta'. Hence, they believed they were 'Web 2.0'. (We say, maybe they have too many bugs; hence, the frequent software releases). As a quick note at this point, 'Perpetual beta' applies to 'data' more than it applies to software. Think 'indexing' when you talk about perpetual beta.. In a nutshell, there is certainly a lot of confusion and many bandwagon seekers out there! The only practical solution is to view the all the principles together.

The seven principles of Web 2.0

So, What is Web 2.0? The long answer is: a service that follows all (or as many as possible) of the seven principles of Web 2.0. We will discuss a simpler definition later. These seven principles are outlined in Tim O'Reilly's original document 19. We discuss them below:

---

18 www.naturalchild.org
19 http://www.oreillynet.com/pub/a/oreilly/tim/news/2005/09/30/what-is-web-20.html

Ajit Jaokar and Tony Fish

The Web as a Platform

Web 2.0 services make the fullest possible use of the Web. The principle of 'The Web as a Platform' encompasses the following concepts:

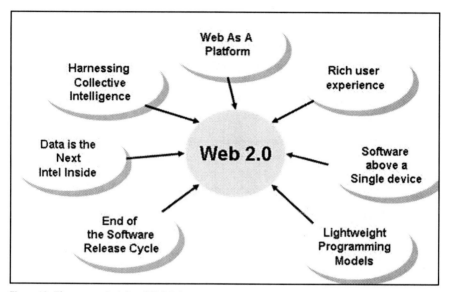

Figure 13: *The seven principles of Web 2.0*

Software as a service:

A Web 2.0 service is a combination of software and data. The term 'web as a platform' is not new. Netscape was an early example of using the Web as a platform. However, Netscape used the Web in context of the existing ecosystem: thus becoming the 'Web Top' instead of the prevailing 'desktop'.

Hence, the usage of the Netscape browser mirrors the famous 'horseless carriage' analogy, wherein an automobile could be described as a 'horseless carriage'. While Netscape was still 'software', in contrast, Google is 'software plus a database', right from the start. (In this context, we use the term 'database' generically to mean that 'Google is managing some data') . Individually, the software and the database are of limited value: but together they create a new type of service. In this context, the value of the software lies in being able to manage the (vast amounts of) data. The better it can do it, the more valuable the software becomes.

Harnessing the 'long tail': The term 'long tail' refers to the vast number of small sites that make up the Web as opposed to the few 'important' sites. This is illustrated by the 'DoubleClick vs. AdSense /Overture' example. The DoubleClick[20] business model was not based on harnessing the vast number of small sites. Instead, it relied on serving the needs of a few large sites (generally dictated by the media/advertising industry). In fact, their business model **actively discouraged** small sites through mechanisms like formal sales contracts. In contrast, anyone can set up an AdSense/Overture account easily. This makes it easier for the vast number of sites (long tail) to use the AdSense/Overture service.

In general, Web 2.0 services are geared to harnessing the power of a 'large number of casual users who often contribute data implicitly' as opposed to 'a small number of users who contribute explicitly'. Tags are an example of implicit contribution. Thus, the Web 2.0 service must be geared to capturing 'many implicit/metadata contributions from a large number of users' and not a small number of contributions from a few 'expert' users.

Harnessing Collective Intelligence

In this context, collective intelligence can mean many things:

- Yahoo as an aggregation of links

- Google PageRank

- Blogging

- Tagging and collective categorisation (for example Flickr[21] and del. icio.us[22])

- eBay buyers and sellers

- Amazon reviews

- Wikipedia

And so on..

---

20 http://www.doubleclick.com/us/
21 http://www.flickr.com
22 http://del.icio.us/

Ajit Jaokar and Tony Fish

All of the above are examples of metadata/content created by users that collectively adds value to the service (which, as we have seen before, is a combination of the software and the data). In addition, harnessing collective intelligence involves understanding some other aspects like **peer production, the wisdom of crowds** and **the network effect**.

Peer production: is defined by the Professor Yochai Benkler's [23] paper, peer production[24].

A concise definition from Wikipedia is:

> *A new model of economic production, different from both markets and firms, in which the creative energy of large numbers of people is coordinated (usually with the aid of the Internet) into large, meaningful projects, largely without traditional hierarchical organization or financial compensation.*

We discuss the motivations underlying peer production in the section on Maslow's laws. The most prominent example of peer production is 'Linux'. However, the same rationale also drives many individuals to contribute in a small way: for example reviews on Amazon. Collectively, these small contributions lay the foundation for the 'Intelligence' of Web 2.0 also called the 'wisdom of crowds'.

The wisdom of crowds: is fully discussed in the book 'Wisdom of Crowds' by James Surowiecki[25]. The central idea of the wisdom of crowds is: large groups of people are smarter than an elite few, no matter how brilliant the elite few may be. The wisdom of crowds is better at solving problems, fostering innovation, coming to wise decisions, and even predicting the future. We discuss the principle of the wisdom of crowds in greater detail in the section on the unified definition of Web 2.0.

And finally, the **network effects from user contributions**. In other words, the ability for users to add value (knowledge) easily and then the ability for their contributions to flow seamlessly across the whole community, thereby

---

23 http://www.benkler.org
24 http://www.benkler.org/CoasesPenguin.html
25 http://www.randomhouse.com/features/wisdomofcrowds

enriching the whole body of knowledge. A collective brain/intelligence of the 'Web' if you will, made possible by mechanisms such as RSS[26].

Data is the Next Intel Inside

Data is the key differentiator between a Web 2.0 service and a non-Web 2.0 service. A Web 2.0 service always combines function (software) and data (which is managed by the software). Thus, Web 2.0 services inevitably have a body of data (Amazon reviews, eBay products and sellers, Google links etc). This is very different to a word processor for example: which comprises of only software (and no data).

While data is valuable, the company need not necessarily own the data. Although in most cases, the company serving the data (for example Google) also 'owns' the data (for example information about links), that may not always be the case. In case of Google Maps [27], Google does not own the data. Mapping data is often owned by companies such as NavTech[28] and satellite imagery data is owned by companies like Digital Globe[29]. Google maps simply combines data from these two sources. Such combination of data from two or more sources is called a **mashup**. A mashup is a website or web application that seamlessly combines content from more than one source into an integrated experience[30]. A mashup could be seen as a 'web API' (Application programming interface).

Taking the 'chain of data' further, sites like Housing Maps[31] are a mashup between Google Maps and Craigslist[32]. The more difficult it is to create the data, the more valuable the data is (for example satellite images are obviously valuable). At the time of writing, there are still grey areas in the ownership and creation of mashups. For instance, if a company makes its data available for 'mashing up', does it control the functionality of ultimate mashup itself? Conflicts could arise if the mashup creator does not use the service in a way approved by the body releasing the API. Nevertheless, there is a significant industry momentum behind mashups.

---

26 http://en.wikipedia.org/wiki/Rss
27 http://www.maps.google.com
28 http://www.navteq.com/
29 http://www.digitalglobe.com/
30 wikipedia
31 http://www.housingmaps.com/
32 http://www.craigslist.com/

Ajit Jaokar and Tony Fish

End of the Software Release Cycle

Web 2.0 services do not have a software release cycle. While Google reindexes its link indices every day, Microsoft releases a major software release every few years. That's because there is no 'data' in Windows 95, Windows XP etc. It's pure software. Not so with Google. Google is data plus software. It has to reindex its 'data' every day, else it loses its value. Thus, operations are critical to a Web 2.0 company and there is no 'software release' as such. The flip side of this coin is, there are widespread beta releases and users are treated as co-developers.

Lightweight Programming Models

Web 2.0 services could be seen as a lighter form of SOA (Service Oriented Architecture). Functionally, a Web 2.0 service acts as a distributed application. Distributed applications have always been complex to design. However, distributed applications are central to the Web. Web services (SOA) were deemed to be the ideal mechanism to create distributed applications easily. But, web services, in their full incarnation using the SOAP[33] stack, are relatively complex. RSS is a simpler (and quicker) way to achieve much of the functionality of web services.

Simpler technologies like RSS and Ajax (explained in detail later) are the driving force behind Web 2.0 services. These lightweight technologies are designed to 'syndicate' rather than 'orchestrate' (orchestration is one of the goals of web services). Because lightweight programming models are oriented towards syndicating data, they are contrary to the traditional corporate mindset of controlling access to data. They are also designed for reuse. In this context, 'reuse' indicates 'reusing the service' and not 'reusing the data' (i.e. they make it easier to remix the service through mashups). As a result of this architecture, grassroots innovation is given a boost because a new service can be created using existing services through mashups.

---

33 http://en.wikipedia.org/wiki/SOAP

Software above the Level of a single device

The sixth principle i.e. 'Software above the Level of a Single Device'. In essence, this book is about the sixth principle. Hence, we do not discuss it in detail here.

A rich user experience

For the Web to be truly useful, we need a mechanism to improve the user experience on the Web. In comparison to platforms such as Windows, the Web offered a relatively limited user experience. Technologies like ActiveX and Java applets attempted to improve the user experience, but these were proprietary. The main technological driver for an enhanced user experience on the Web is Ajax. Ajax uses web technologies (i.e. non-proprietary technologies). Ajax was outlined by Jesse James Garrett on an article published on the Web34. Ajax is being used in services like Gmail, Google Maps and Flickr and it already provides the technology to create a seamless user experience combing many discrete services.

We will discuss Ajax and the enhanced user experience in subsequent sections.

Web 2.0 – A unified view

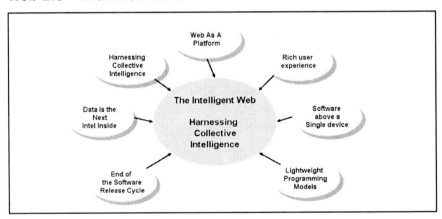

Figure 13: *The intelligent web*

---

34 http://www.adaptivepath.com/publications/essays/archives/000385.php

Ajit Jaokar and Tony Fish

Having discussed the seven principles of Web 2.0, let us now look at a simpler definition of Web 2.0. We can all this definition as the 'unified view' of Web 2.0.

If we reconsider the seven principles, we observe that the second principle (harnessing collective intelligence) encompasses the other six.

Thus, we can view Web 2.0 as '**Harnessing Collective Intelligence**' or '**The Intelligent Web**'. What kind of intelligence can be attributed to Web 2.0? How is it different from Web 1.0?

Web 1.0 was hijacked by the marketers, advertisers and the people who wanted to stuff canned content down our throat! The dot com bubble was the end of many who took the approach of broadcast content. What's left is the Web as it was originally meant to be a global means of communication.

The intelligence attributed to the Web (Web 2.0) arises from **us** (i.e. the collective/people) as we begin to communicate.

Thus, when we talk of the 'Intelligent web' or 'harnessing collective intelligence': we are talking of the familiar principle of the 'wisdom of crowds'. In order to harness collective intelligence.

a) Information must flow freely

b) It must be harnessed/processed in some way - else it remains a collection of opinions and not knowledge

c) From a commercial standpoint, there must be a way to monetise the 'long tail'

Our essential argument is: *if we consider Web 2.0 as 'Intelligent web' or 'Harnessing collective intelligence' (Principle two), and then look at the other six principles feeding into it, Web 2.0 is a lot clearer*

Since the wisdom of crowds is so important - let's consider that in a bit more detail. From the Wikipedia entry for the wisdom of crowds [35]

Are all crowds wise? No. They are not.

The four elements required to form a 'wise' crowd are:

a) Diversity of opinion.

b) Independence: Peoples' opinions aren't determined by the opinions of those around them.

c) Decentralisation: People are able to specialise and draw on local knowledge.

d) Aggregation: Some mechanism exists for turning private judgements into a collective decision.

Conversely, the wisdom of crowds fails when:

a) Decision making is too centralised: The Columbia shuttle disaster occurred because the hierarchical management at NASA was closed to the wisdom of low-level engineers.

b) Decision making is too divided: The U.S. Intelligence community failed to prevent the September 11, 2001 attacks partly because information held by one subdivision was not accessible by another.

c) Decision making is imitative - choices are visible and there are a few strong decision makers who in effect, influence the crowd.

---

35 http://en.wikipedia.org/wiki/Wisdom_of_crowds

Ajit Jaokar and Tony Fish

Now, let's look at the seven principles again:

1. The Web as a Platform

The Web is the only true link that unites us all together whoever or wherever we are in the world. Hence, to harness collective intelligence and to create the intelligent web, we need to include as many people as we can. The only way we can do this is to treat the Web as a platform and use open standards. You can't harness collective intelligence using the IBM ESA/390[36], no matter how powerful it is!

2. Harnessing Collective Intelligence

Now becomes the 'main' principle or the first principle

3. Data is the Next Intel Inside

By definition, to harness collective intelligence, we must have the capacity to process massive amounts of data. Hence, data is the 'intelligence' (Intel)

4. End of the Software Release Cycle

This pertains to 'Software as a service'. Software as a 'product' can never keep upto date with all the changing information. A Web 2.0 service includes code as well as data. Thus, 'software as a service' keeps the data relevant (and the harnessed decision accurate) by accessing as many sources as possible

5. Lightweight Programming Models

The heavy weight programming models catered for the few. In contrast, using lightweight programming models, we can reach many more people. Hence, we are working with many more sources of information, leading to an intelligent web. For example: from the seven principles[37]

---

36 http://en.wikipedia.org/wiki/System/390
37 http://www.oreillynet.com/pub/a/oreilly/tim/news/2005/09/30/what-is-web-20.html?page=1

Amazon.com's web services are provided in two forms: one adhering to the formalisms of the SOAP (Simple Object Access Protocol) web services stack, the other simply providing XML data over HTTP, in a lightweight approach sometimes referred to as REST (Representational State Transfer). While high value B2B connections (like those between Amazon and retail partners like ToysRUs) use the SOAP stack, Amazon reports that 95% of the usage is of the lightweight REST service.

6. Software Above the Level of a Single Device

More devices to capture information and better flow of information between these devices leads to a higher degree of collective intelligence.

7. Rich User Experiences

A rich user experience is necessary to enable better web applications leading to more web usage and better information flow on the Web: leading to a more 'Intelligent' Web.

Web 2.0 – A mini FAQ

To summarise, here is a mini FAQ:

What is Web 2.0? It's the intelligent Web (the second principle: harnessing collective intelligence)

What makes it intelligent? We do.

How does it happen? By harnessing collective intelligence

What do you need to harness collective intelligence? The other six principles!

Mobile Web 2.0 – An introduction

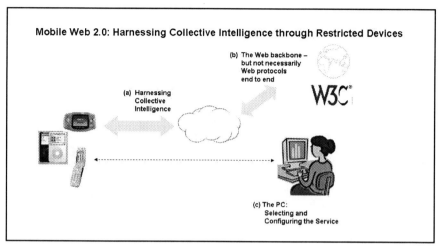

Figure 14: *What is Mobile Web 2.0*

Having discussed Web 2.0, let us consider the implications of extending the definitions of Web 2.0 to Mobile Web 2.0. As we have seen previously:

> Mobile Web 2.0 is focused on the user as the creator and consumer of content 'at the point of inspiration' and the mobile device as the means to harness collective intelligence

Thus, Mobile Web 2.0 extends the principle of 'harnessing collective intelligence' to restricted devices. The seemingly simple idea of extending Web 2.0 to Mobile Web 2.0Web 2.0 has many facets, for instance:

a) What is a restricted device?

b) What are the implications of extending the Web to restricted devices?

c) As devices become creators and not mere consumers of information – what categories of intelligence can be captured/harnessed from restricted devices?

d) What is the impact for services as devices start using the Web as a massive information repository and the PC as a local cache where services can be configured?

Restricted devices

A broad definition of a 'restricted device' is not easy. The only thing they all have in common is - 'they are battery driven'. But then, watches have batteries? A better definition of restricted devices can be formulated by incorporating Barbara Ballard's carry principle 38, which introduces the concept of an information device. Using the carry principle, a restricted device could now be deemed as:

a) Carried by the user

b) Battery driven

c) Small(by definition)

d) Probably multifunctional but with a primary focus

e) A device with limited input mechanisms(small keyboard)

f) Personal and personalised BUT

g) Not wearable (that rules out the watch!). But, there is a caveat, a mobile device in the future could be wearable and it's capacities may well be beyond what we imagine today. The input mechanisms in the future may not be keyboards. So, this is an evolving definition.

Finally, there are differences between a small screen **'carry/portable'** device such as mobile phones and a 'device which is carried', such as in plane entertainment, which we call a **'carried/ported'** device. For example, in a car, a GPS navigator is a 'small screen based mobile device' and in a plane, the in-flight entertainment screen is also 'mobile'. However, both these devices are not 'carried by a person' and do not have the same screen/power restrictions

---

38 http://www.littlespringsdesign.com/blog/2005/09/14/the-carry-principle/

Ajit Jaokar and Tony Fish

as devices that people carry. However, whichever way you look at it, it's clear that the mobile phone is an example of a restricted device. From now on, we use the definition of mobile devices interchangeably with 'restricted devices' and the meaning will be clearer in the context.

Extending the Web to restricted devices

It may seem obvious, but Web 2.0 is all about the 'web' because Web 2.0 could not have been possible without the Web. Thus, in a 'pure' definition, Web 2.0 is about 'harnessing collective intelligence via the Web'. When we extend this definition to 'Mobile Web 2.0' – there are two implications:

a)   The Web does not necessarily extend to mobile devices.

b)   Even though the Web does not extend to mobile devices, intelligence can still be captured from mobile devices.

The seven principles of Web 2.0 speak of this accurately when they discuss the example of the iPod/iTunes. The iPod uses the Web as a back end and the PC as a local cache. In this sense, the service is 'driven by the Web and configured at the PC' but it is not strictly a 'Web' application because it is not driven by web protocols end to end (iPod /iTunes internals are proprietary to Apple).

Thus, the characteristics (distinguishing principles) of Mobile Web 2.0 are:

a) Harnessing collective intelligence through restricted devices i.e. a two way flow where people carrying devices become reporters rather than mere consumers

b) Driven by the Web backbone, but not necessarily based on Web protocols end to end

c) Use of the PC as a local cache/configuration mechanism where the service will be selected and configured

The Seven Principles Of Mobile Web 2.0

Introduction

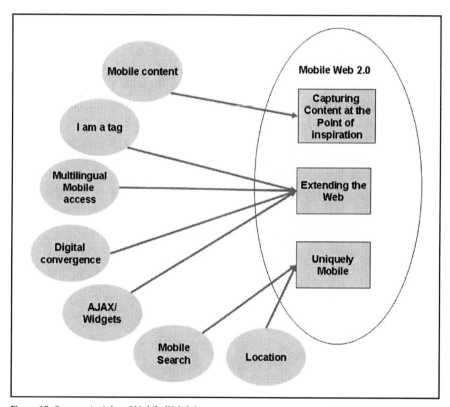

Figure 15: *Seven principles of Mobile Web 2.0*

Like Web 2.0, Mobile Web 2.0 can also be outlined in seven principles. Because Mobile Web 2.0 exists within the definition of Web 2.0 itself, we use the following three guidelines to define the seven principles of Mobile Web 2.0.

a) Harnessing collective intelligence

b) Capturing information at the point of inspiration

c) Extending the Web

The seven principles of Mobile Web 2.0 are:

1) Content created on the mobile device will change the balance of power in the media industry.

2) I am a tag, I am not a number: Tags could provide a way to map the various 'numbers' in our life. This frees us from the restrictions of network operators(both fixed and mobile)

3) Multilingual mobile access will be critical to Mobile Web 2.0: Mobile Web 2.0 will be multilingual. It will impact many more people whose first language is not English and who reside in developing countries.

4) Mobile Web 2.0 will drive digital convergence.

5) Ajax is a disruptive force within Mobile Web 2.0

6) Mobile Web 2.0 will drive location-based services

7) Mobile Web 2.0 will drive mobile search

In the following sections, we discuss these seven principles in detail.

Mobile content and the changing balance of power
Synopsis

Content created on the mobile device will change the balance of power in the media industry. The mobile device is ideally poised to capture user generated content 'at the point of inspiration' – making it the main driver behind Web 2.0. This new role is affecting the role of other 'screens' in our life (such as the PC and the TV) and is triggering a change in the balance of power in the content creation industry.

To explain these concepts, we have to understand a range of ideas that are at the boundary of technology and user behaviour. In the following discussion, we shall first present the importance and significance of user generated content and then discuss the impact of mobile devices on the value chain.

The six screens of life and the existing flow of content

For the most part, we are consumers of content. In our daily lives, we consume professionally created, produced and edited content from traditional and new media providers on our "six screens of life". These "six screens of life" are divided into two broad categories, big screens and small screens, each with three sub groups.

The 'BIG' screens of life

Cinema (shared with other members of the public):
TV (shared privately within our homes)
PC (personal or shared use)
The 'small' screens of life

Fixed/Portable Players (fixed devices in things that move such as cars, planes, etc)
Information screens e.g. iPod, radio.

Both for Big and Small screens, the user has traditionally been a passive receiver of content (i.e. content has been broadcast to the user) OR the user has been seen as a member of a carefully controlled and managed audience (for example voting) – **but not as a primary creator of content**.

For instance:

- The TV needs users to consume(view)

- A website needs users to consume/interact in most cases

- Cinema needs users to consume

Reflecting the above trend, most of the content on the Mobile device to date, has also been the 're-presentation' and 're-production' of existing material delivered to the mobile screen. **This chapter will discuss how the mobile device is changing from being a primary consumer to a major creator of content.** The six screens of life and the creative/consumption balance can be seen below.

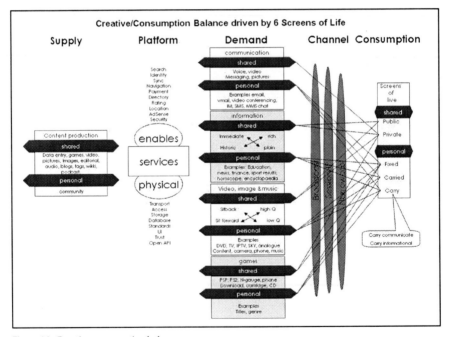

Figure 16: *Creative consumption balance*

Figure 16 is divided into five broad areas. The three main areas are: **Supply, Demand and Consumption**. These are supplemented by the Platform and the Channel.

The supply side

The supply side of the diagram details content production. Some content is created by professional media companies (for example: movies, music), and other content by the individual (for example: blogs, wikis). Eventually, the content is consumed in a private or a shared fashion.

The platform

Content created from the supply sources is passed through a platform. The platform has two components of service: - the physical components and the enabling components.

Examples of the physical components are:

- **The transport network:** which provides a means to carry the data from production to the point of access.

- **The access layer:** that allows the user to connect and receive the content.

- **Storage:** the component that holds the content until it can be consumed.

- **The database:** that makes content available for searching.

Examples of the enablers are:

- **Search** – the engine that drives the ability to search

- **Identity** – who people are

- **Sync** – synchronisation to ensure quality of service

- **Navigation** – enabling ease of use

- **Payment services** - to ensure that the economics work

- **Directory** – ability to find

- **Rating** – what is good and what is not good

- **Location** – A mapping of location and content

- **AdSense** – ability to serve ads within content

- **Security** – protection

The platform is built by companies who want to take content and deliver it to customers. Each company will offer different physical and enabling components depending on their business models.

Demand

Demand is classified into a number of services, each broken into shared and personal modes of consumption. These services include:

- **Communication:** Shared services such as video and pictures and personal such as IM, SMS and email.

- **Information services:** Information is a matrix of immediate or historic delivered in either a rich or a plain format. Examples of information services could include information delivered on the TV in a rich media format and the same information would be delivered on the mobile browser in a text format.

- **Video, Image and Music:** Like information, this group is a matrix, depending on sit forward or sit back, and adjustments for the quality of the service needed. Examples would be CD quality vs. MPEG3; depends on listening in the living room with $4,000 HiFi vs. an iPod on the train.

Ajit Jaokar and Tony Fish

- **Games:** Games and gaming is a complex area and is partly driven by the platforms but also by titles. However, like other demand services, it splits into personal and shared for playing.

Channels

The channels of delivery are broadly broadcast, session IP and narrowcast. **Broadcast** represents TV and other broadcast channels. **Session/ IP** represents the Internet and **narrowcast** represents the mobile device. Unlike Broadcast, narrowcast has fundamental limitations due to the physics of the device itself, battery life and radio propagation. The model (supply – platform – demand – channel and consumption) ends with the content created being consumed on a variety of screens on a shared or a personal basis.

Consumption

Consumption is based on the six screens of life. To recap, these are:

The 'Big' screens of life

- Cinema (shared with other members of the public):

- TV (shared privately within our homes)

- PC (personal or shared use)

The 'small' screens of life

- Fixed/Portable Players (fixed devices in things that move cars, planes, etc)

- Information screens e.g. iPod, radio.

- Mobile-communicator, an individual and personalised handheld device

These screens could be classified into: Fixed, Carried and Carry screens.

Fixed: Fixed screens are physically confined to a space for example the TV, PC.

Carried: A carried device is a device which could be on the move BUT is not personal. In flight entertainment is an example of a carried device.

Carry devices: Carry devices are portable, lightweight and free of wires, the prime example being a mobile phone

Further, the devices could be classified into:

a)   **Shared** : where more than one person could view the screen and

b)   **Personal** : where one screen corresponds to one viewer

We make two observations with respect to consumption patterns:

Consumption is becoming increasingly personal especially with mobile (carried) devices and

Private consumption needs an understanding of the context and the content. Merely transforming content from the shared, bigger screens to private, smaller screens is not enough. We explore the idea of context in subsequent sections in detail. Specifically, in relation to mobile services, we note that existing mobile services are relatively simple and don't take into account the content or the context. Like the Internet, mobility is all about communications. The end user is the 'killer application' providing both the content and the context to services such as voice, email, txt and IM. When we initiate a conversation (contact) we take into account the context of the message – for example, who we are communicating with (Partner, Work Colleague, Friend, Financial Advisor, Utility Service Provider, Customer, etc). This context also drives the content of the message (what we are actually saying). We also take into consideration other parameters, such as the time of day, time we have, the level of urgency, emotional state, location, the bill payer, etc.

Ajit Jaokar and Tony Fish

So far, the network operators, fixed and mobile, have been successful in provisioning these en-mass communications services. These services lend themselves to provisioning on a large scale and require broad based (un-targeted) marketing to their customer base. They are not complex and inherently price driven. As communications markets mature (both in penetration and in reach), new value will be derived from applications which lower costs for the end user, leverage the contextual and situational aspects and/or exploit disruptive technologies.

Content is King – unless the creator changes

The content value chain as depicted earlier, is oriented towards broadcast content. True, there are some user generated components in that chain – such as blogs etc, but the value chain itself is designed for pre packaged broadcast content. Hence, as we have seen previously, the prevailing industry notion amongst members of the broadcast content industry is 'Content is King', which runs contradictory to our thinking outlined previously in this book.

However, the notion of '(Broadcast) Content is King' changes dramatically when user generated content becomes the dominant content type.

In this section, we introduce three ideas:

a)   Content creation is triggered by events

b)   User generated content is increasingly being consumed by the community and

c)   The community could take on some of the functions of the editor.

To understand these concepts, let us consider the basic nature of content itself. Content is created due to the occurrence of an **event**. No matter how it is triggered, content falls broadly into four value brackets:

- **Unique content**: such as a film epic whose value increases or decreases depending on marketing and fashion (trends). Unique content could also include a piece of art work or innovation, whose value increases with each passing year.

- **A body of work**: such as a database that could be used for statistical analysis.

- **New and News**: A new story or a re-release of something that has come back into fashion. In just about all cases, news value falls off within 24 hours for there is something newer. Its value is short and time limited.

- **Brand**: in rare cases, an event will become a brand, such as Live8[39], which, if it reaches unique status, produces increasing value over time.

There are a wide number of commercial techniques that can be employed for the exploitation of content generated by these events; which can include the sale or licensing of rights for use. The content thus produced was originally consumed only in the mass market media. Now, that content is increasingly being consumed by the 'community'. By community, we mean people who are no longer passive consumers but are influenced by direct recommendations from friends they know, ratings from peers they don't know and statistical aggregations (such as Google PageRank or Alexa ratings).

---

39 http://www.live8live.com/

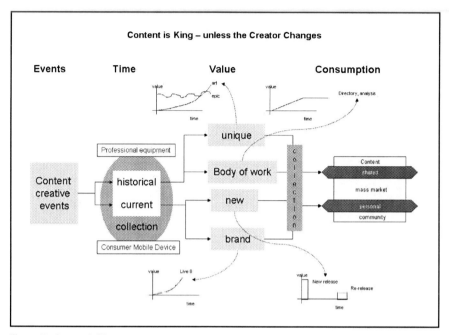

Figure 17: *Content is king*

The role of the community also impacts on the role of the editor. In the old media economy, the editor was the person who had an idea. In other words, the content creation process started with the journalist/editor. While the editor was not often the expert, the editor was certainly knowledgeable about the subject and had a passion for the content. The business model for the final content produced was simple: either sponsorship (advertisement) or price (subscription) or a mix of both. The editor had now created a brand i.e. the publication. Over time the editor moved on, and a new editor was selected for their insight and understanding of the segment. Taking on from the original editor, this new editor's role was to build the brand further or maintain its values. As new editors came and went, slowly, passion was replaced by 'following the brand'/conforming to the 'house style'. Assuming that the publication/production worked i.e. made money: this formula becomes self reinforcing.

But, success had a price! Ironically, the brand, which the editor created out of his own passion, had become a shackle. The brand and readership move in step and create a symbiotic relationship. The editor's viewpoints are predictable and the readers are used to a viewpoint within a certain frame of reference (for example: many newspapers lean either to the left or to the right of the political spectrum. Their 'news' is often tainted by the makeup of their political 'lens'). Around the rise of the Web (late 1990s/2000), this cosy relationship changed forever.

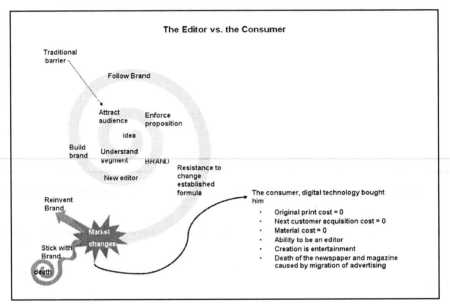

Figure 18: *Editor vs. the consumer*

The consumer became the creator. The traditional barriers of print cost, customer acquisition have migrated to zero. Everyone is now empowered and takes on the role of a creator. At first, the media ignored the threat of user generated content, as they still had a focus on embracing new digital channels. But, when advertising dollars began to migrate to newer online properties and communities such as MySpace the industry was hit hard.

Is there a role for the editor? In some ways, there is, because as the sheer volume of content grows, there is a need to select the good from the bad.

Here, the community could step in to perform the function of the editor (the filter). We discussed previously the examples of Digg[40] and YouTube[41] where the community rates content and thereby acts as a filter. Hence, user generated content and communities are complimentary because they are two facets of the same equation.

User generated content – changing the business models

In the previous section, we have seen how user generated content is changing the balance of power. Now, we will see how business models are being affected by user generated content.

User generated content, as presented in the value eco-cycle, is generated from life events. The inputs triggering user generated content are either from the highlights or from the 'low-lights' of a continuous recording (similar to the TV show 'Big Brother'[42]) OR from a specific stimulus, such as entertainment, a thought, news or an idea. This is captured pictorially in the figure below:

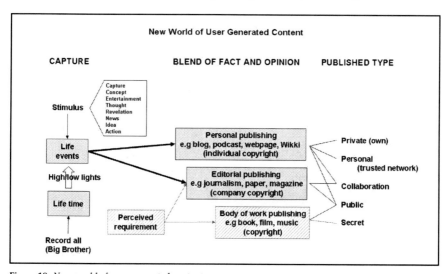

Figure 19: *New world of user generated content*

---

40   www.digg.com
41   www.youtube.com
42   http://www.channel4.com/bigbrother/

From these life events will arise either personally published content (such as blogs) or professionally published content (such as magazines). Note that the distinction in this case pertains to ownership of copyright rather than the quality of the content itself. Besides personally published content and professionally published content, there is a third (less common) way of triggering content. This is driven from a different stimulus, which we call the "perceived requirement". An example of this mode could be a consulting project or a public enquiry into a specific incident of public interest, or a secret requirement such as understanding the impact of alternative energy.

Personal and editorial publishing lead to similar but different publishing end goals, irrespective of whether the content is audio, text or video. Personal types of publishing are:

- **Private (own) use:**
  In this case, the information is shared with no one else. Examples include an online anniversary card/message.

- **Personal use:**
  In this case, the information would be shared with a trusted network – for example: personal photos.

- **Collaboration use:**
  In this case, the user is submitting their opinion to a Wiki or other similar collaborative service.

- **Public use:**
  In this case, the user may have captured some newsworthy event that is published on a site for public access and consumption under a creative commons license.

In a similar way, editorial publishing would use both collaborative and public publishing modes; but with the copyright resting with the company, rather than the individual. We believe that the worlds of user generated content and editorial content will not co-exist until 2012 in a mutually consistent economic balance as shown in figure 20.

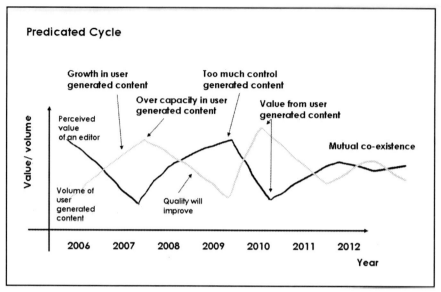

Figure 20: *Predicated cycle*

We predict that the volume of user generated content will continue to grow well into 2007, however, we believe that there is likely to be an oversupply of rather poor quality material that will lead to the return of power to the editor in 2008. At this stage, the editor will pick and select the best of both user generated content and also editorially produced. To some extent, this is already happening with some large publications accepting content from the public, then editing or aggregating it in some way. The community itself can edit to some extent of course; but an editor is required especially if the material needs to be presented in a summarised or a concise form. In other words, an editor performs two functions: selecting what to print and then presenting that content in a concise/readable format. The first function can be performed by the community but for the second function, we still need an editor. Thus, editors will recover their content control status after an initial erosion of power.

However, these new editors will overtly control, hence giving rise to a new wave of user generated content in 2009/ 2010, driven by richer content such as video. Media such as video and podcasting do not have the same requirement of needing to be presented in a concise form because they are logically complete and they can be skipped to watch the next clip. For instance, in the UK, you can view 'You've been framed'[43] clips in isolation. The same applies to a number of situations like comedy video clips, animation video clips etc. It could well be 2011/2012 before a balance is reached between editorial content and user generated.

We believe that this is the same time period needed to change the broken adverting business models to new metadata and attention driven business concepts. Users will repel the old business models of advertising, which give rise to the question of: 'Who pays?' Given that we perceive that users don't value editorial content sufficiently on the Web to pay for it as users do for content in print; a new economic balance will not be reached until 2012.

TV, Radio and digital communities will all suffer. The 'new model' will be predicated on the ability to collect information on what the user is doing; this collected information will be in the form of meta-data. This meta-data will be collected automatically (for instance time, attention and location on a mobile device). Mobile will be critically important; as time spent by a consumer at a fixed web access node will represent an ever decreasing percentage of their total web interactions.

Meta-data gathered from fixed access alone will become an ever-less important predicator of consumer requirements. Mobile devices however, will enable data to be collected and analysed 'at the point of inspiration' or more importantly 'at the point of need'. In the new model, with collected metadata, consumers will be offered services on demand; e.g. insurance will have a 'cover period' that varies by activity. Therefore, value will be generated in the connection between creation and consumption via relevance, voting, index, tags, search, verification, attention and payment as shown in figure 21.

---

43 http://www.metacafe.com/watch/72947/youve_been_framed_outdoors/

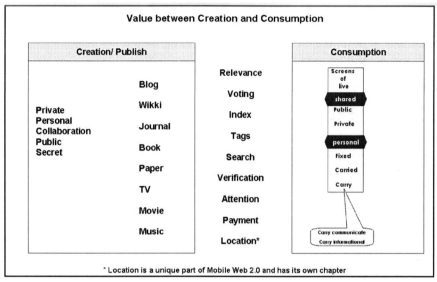

Figure 21: *Value creating components between creation and consumption*

The value creating components between creation and consumption are:

Relevance: [44]

This is a feature that creates an output based on history. The output is formed from the marriage of the new input requirement and the previous meta-data history of the individual. The meta-data is information such as the 'clickstream and attention. There is an implicit assumption here, that the past is a good indicator of the future. Relevance as a function allows the provider of the service to predict requirements based on history.

Voting: [45]

Through voting, users provide feedback (something purchased) or 'feedin' (voting on Big Brother/ Pop Idol) on services and products that the consumer is interested in. Voting is used to provide recommendation and/ or rating/ reputation, such as those employed by Amazon and eBay.

---

44 http://en.wikipedia.org/wiki/Relevance
45 http://en.wikipedia.org/wiki/Voting

Indexing: [46]

Indexing is designed to help the user find information quickly and easily. Indexing, in this sense, is a well-organised map of Web content, including cross-references, grouping of like concepts, and other useful intellectual analysis.

Tags: [47]

This is the data that a user adds to define what a certain piece of content is about, this is partially useful for finding audio, pictures and video content and summaries of text and will be discussed later in the book, in detail.

Search: [48]

A program that helps to search information from the Web.

Verification:

This is the establishment of the veracity (truth) of something. In the context of Mobile Web 2.0 is being used as part of the verification of a user, i.e. the user is who they say they are. What is unique here is that verification of an individual comes from the social or trusted group and not from government papers.

Attention:

In its widest sense is a cognitive process of selectively concentrating on one thing while ignoring other things. In the context of Mobile Web 2.0, attention refers to the ability to gather information on what the user has given attention too. The information is gathered from the users activities; for instance the time users spend with a particular function/screen.

Payment:

This is the act of transferring wealth to another person or company. Why this is of value is usually self-evident!

The level of the value created by connecting components described above varies. While they can be recognised as individual components, they all

---

46 http://en.wikipedia.org/wiki/Indexing
47 http://en.wikipedia.org/wiki/tags
48 http://en.wikipedia.org/wiki/Search_engine

require interoperability of systems and development of software before they can be fully exploited. Much work has been published on many of these subjects, *and we believe that voting, index, search and payment are well developed however relevance, tags, attention and verification are weak.* However, relevance, tags, attention and verification are the key components in the shift to the new business models; as these will form the core of the new class of 'meta-data'. We believe that these topic areas and subsequent developments will be of keen interest to the dominant web companies and venture capital alike.

To bring some of these concepts together; in the future, we will all be creators as well as consumers; this could be as simple as providing feedback/ voting [raters and reviewers] on existing sites. Peer review systems exist in closed environments today, the obvious ones being eBay reputation and Amazon review. But, in the future, these will be widespread.

However, in the future 'Web 2.0' will build a distributed rating and review system, such that if I review your blog or someone's record or video; then instead of placing it on their site, I will add the comment on "MySpace" and then ping the originator with my thoughts, an extension of trackback in blogging for products and services. 'Pingerati' to coin a new word, could be an example. By accumulating all of my comments, content and reviews on one site I can then build a reputation, as peers and individuals review my collective thoughts. If I tag and review on one area in particular, I could become considered as an expert in that field, as long as my reputation supports this. Hence, my peers will then ultimately determine if my reviews and opinions are of value. Taking this thought one stage further, the system (Web 2.0) would allow these peers who rate my comments to provide a validation of who I am. Therefore my verified Identity comes, not from official paper work, but by my actions and views. This verified identity also has no links to my bank account (as it is about created content on the Web) and therefore has less value to identity fraud, but could be used as a method to buy simple services, these thoughts are explored later.

The changing creation/consumption balance

In the previous sections, we introduced the following ideas:

a) The six screens of life

b) Content creation is increasingly becoming personal. Content and context are important

c) Content creation is triggered by events

d) Content is increasingly being consumed by the community

e) The community could take on some functions of the editor.

f) User generated content, changing the business models.

In this section, we see how the balance between creation and consumption of content is changing. We will introduce two more ideas:

a) The creative / consumption balance was **originally** driven by demand

b) The creative / consumption balance is **now** driven by supply

This ultimately leads to a new value chain, which favours user generated content. Consumption of content is driven by the need to entertain or to inform. This need (demand) was the driving point of the old value chain. In the old media world, the consumers (business or individual), request (demand) content from the service providers to satisfy their needs for entertainment or information.

The service providers were strong brands (Bloomberg, Sky, Reuters, FT etc). These brands, in turn, would employ creators of content (in some instances, other businesses) to gather and create the content. However, they would retain the rights of ownership, unlike the true freelance creators. These creators of content (including freelance) would deliver the content to a platform (via the aggregating 'brand'), which would publish and supply the content to the consumer. A happy balance was achieved as the demand and supply could be met for a fair economic return.

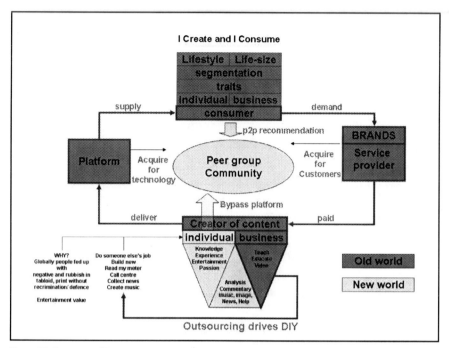

Figure 22: *I create and I consume*

Over time, under pressure from the city and other financial motivations, the trend towards outsourcing (freelance) increased. Meanwhile, a second group of 'freelancers' were emerging. For businesses as a whole in all sectors, there is now a trend towards greater self employment, freelancing, contracting etc.

This second group of freelancers (i.e. the outsourced employee), provided a ready pool of talented, knowledgeable people. Thus, a steady talent was being built 'outside the corporate walls' comprising of an army of freelance journalists and freelancers from the industry itself. The Web (specifically blogging) and broadband simply provided the vehicle for this talent to popularise their views. It also enabled these new independent individuals to publish their own content, without having to revert to the old media world for distribution.

Soon a virtuous circle emerged, where it became increasingly easy to generate content (creation) and publish it. With zero cost of materials and access to a wide readership, new creators of content emerged. This led to a **supply led value**

chain (to be contrasted with the existing **demand led content value chain**) supplemented by concepts such as P2P (Peer to peer), communities etc.

The motivations in the previous, demand led, value chain are different from those in the new, supply led value chain. Previously, the demand arose **first** from the customers and was driven by a specific need. Now, **the content is first created by the suppliers** who are motivated by 'self actualization' needs as depicted by Maslow's hierarchy of needs. Briefly, Maslow's theory of hierarchy of needs could be depicted as below.

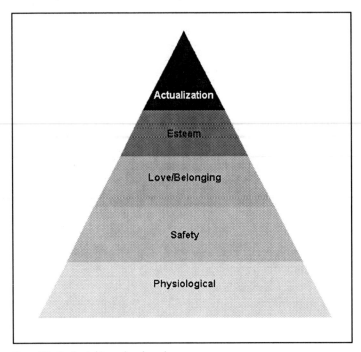

Figure 23: *Maslow's hierarchy of needs.*

1. Physiological
2. Safety
3. Love/Belonging
4. Esteem
5. Actualization

According to Maslow, the needs emerge incrementally i.e. when the lowest needs are fulfilled, the organism looks to fulfil higher needs. Physiological needs are the most basic (lowest) needs of an organism. These include food and sleep. Next are the safety needs; these include physical security, security of employment etc. When these needs are satisfied, the individual seeks love and belonging which include social interaction, friendship and the need to belong to a group/association/religion etc. Next, the individual seeks esteem and respect from their community and peers.

Finally, when all these needs are satisfied, the individual seeks 'self actualization. Self-actualization (a term created by Kurt Goldstein) is the fundamental need of human beings to make the most of their talents (unique abilities). Maslow described it as follows: *Self Actualization is the intrinsic growth of what is already in the organism, or more accurately, of what the organism is. (Psychological Review, 1949)*

Maslow writes the following of self-actualizing people:

- They embrace the facts and realities of the world (including themselves) rather than denying or avoiding them.
- They are spontaneous in their ideas and actions.
- They are creative.
- They are interested in solving problems; this often includes the problems of others. Solving these problems is often a key focus in their lives.
- They feel a closeness to other people, and generally appreciate life.
- They have a system of morality that is fully internalized and independent of external authority.

- They judge others without prejudice, in a way that can be termed objective.

References to Maslow. source: Wikipedia[49]

Thus, self actualization is the need that motivates the bloggers and others who create content. *Ironically, this need is similar to the original editors (passion for their jobs and their subject matters) – before they were subsumed by commercial considerations and the brand.*

Thus, the creation process has come full circle!

A new value chain emerges

So far, we have covered a range of ideas that build upon each other.

Let's recap them once again.

In the previous sections, we introduced the following ideas:

a) The six screens of life

b) Content creation is increasingly becoming personal. Content and context are important

c) Content creation is triggered by events

d) Content is increasingly being consumed by the community

e) The community could take on some functions of the editor.

f) The creative / consumption balance was originally driven by demand (entertainment/information)

g) User generated content, changing the business models

h) The creative / consumption balance is now driven by supply (Maslow).

The existing value chain has limitations. Although it includes some user generated content, it is not suited for the New World of content which is **predominantly**

---

49 http://en.wikipedia.org/wiki/Maslow's_hierarchy_of_needs

user created. This is because, the same basic building blocks are utilised in both cases, but in different directions. ***Hence, we have a loop instead of a chain.*** The entry point is the same for creation and consumption i.e. an event.

However..

- The event for personal creation is driven by a "point of inspiration".

- The event for content consumption is driven by a "point of entertainment/ information or interaction".

In the value loop (figure 24), if we go from left to right: the user creates content and publishes it. If we go from right to left, content is created for the user to consume. In both directions; the same/ identical device and access infrastructure will be utilised **but the content and its context is different.**

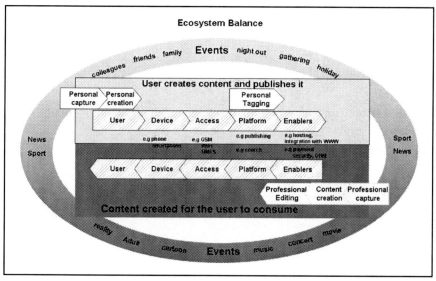

Figure 24: *Ecosystem balance*

To explain this, let's approach the same 'event' from the left (user perspective) and the right (professional media perspective). The event could be a sporting event, for example a soccer match.

If the user records the event, the user captures the event on their mobile device and the content is uploaded via the access network to a platform. The platform then enables the user to publish and edit the content maybe on the device or via a PC. **Note that, in this process, the content had metadata about location and time, and also any additional tags (annotations) made by the individual who recorded it.** This metadata allows for easy searching and indexing. It is much easier to find the same data by location than by an individual's description of that picture.

In contrast, moving from right to left (professionals record the event), the same event is edited and delivered to the same user as content to be consumed. In this case, it is a push model, but content will become increasingly hard to find.

Both models record the same event and both models will co-exist.

One could argue that the professionals produce content of better quality. However, there is no doubt that the impact of new technology is being felt.

This is evident from the BBC[50], The FT [51], CNN[52] and the Wall Street Journal[53]. All of whom are moving forward with a wide range of new media/technology solutions.

Finally, note that as the events move about the top of the diagram, they become increasingly more personal. In the case of personal events (which the users value most of all), the user is indeed the only source of information.

---

50 www.bbc.com
51 www.ft.com
52 www.cnn.com
53 www.wsj.com

Ajit Jaokar and Tony Fish

Significance of the mobile device – Maslow revisited

So far, we have been discussing user generated content in general. Let us now look at the mobile device based on this background. The mobile device occupies a unique place in our lives due to its personal nature. As it grows more powerful in terms of processing, memory, bandwidth and connectivity; it can fulfil higher needs for us. Again, Maslow comes to mind, as depicted below.

In context of Maslow's needs, The basic physiological needs correspond to basic services like voice and text. The safety and security needs correspond to being able to contact people when you are lost, child tracking etc. The love and belonging needs could translate into mobile dating, flirting and community services. The esteem needs can be seen in people who personalise their mobile devices. Finally, the self actualisation needs are at the pinnacle of our hierarchy and are now being met in creation of content.

This relationship is unique i.e. one does not feel the same emotional empathy with their PC as they do to their mobile device.

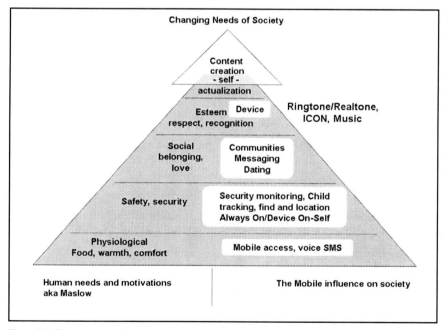

Figure 25: *Changing needs of society*

Mobile content and the changing balance of power

Finally, in this section, we shall bring all these ideas together.

We have discussed:

- The six screens of life

- Content creation is increasingly becoming personal. Content and context are important

- Content creation is triggered by events

- Content is increasingly being consumed by the community

- The community could take on some functions of the editor.

- Business models are being affected by user generated content

- The creative / consumption balance was originally driven by demand

- The creative / consumption balance is now driven by supply (Maslow)

- These ideas lead to a new value chain oriented to user generated content (which is more a loop than a value chain) and includes both content and context

- Finally, the mobile device is central to our life. Our emotional involvement with the mobile device touches many levels (la Maslow)

Thus, if we accept the importance of user generated content, then the balance between the screens changes (and consequently the balance of power between the players).

Mobile Web 2.0 devices will drive the capture of inspiration ranging from pictures, calendars (the things you need to remember), notes and reminders (viewpoints and ideas that you want to work on later), news (citizen's reporting) etc. These will be then stored in your private web space. This information will include explicit tags (user annotations) and also implicit tags (like location information).

PC 2.0 will provide powerful tools for editorial, configuration and cache (local storage).

TV 2.0 will be the primary point of viewing all these services.

This holistic approach will drive a converged media experience.   Creation will be driven by the mobile device and consumption by the PC/TV. All three screens are needed to provide a balance.

Is this theory or is there factual instances of the power of user generated content?

Back in the 2002 World Cup, we were all still debating about the success of 3G and broadband was still not a major factor.

Today in 2006, FIFA (the world soccer governing body) has a very different problem on hand. Initially, FIFA stipulated that no pictures of games should appear on websites until the final whistle and that these pictures should be limited to five per half.[54] Under pressure from the world association of newspapers and the sponsors, FIFA changed its mind.

But, what about the millions of people who could send pictures 'live' from the match directly to their own blogs and other sites?

After all, it's very easy to do so using a site like http://moblog.co.uk/

FIFA cannot control them all! The fans at the live event could be truly 'broadcasting' from within the stadium. They are thus competing with the big media!

---

54 http://www.pressgazette.co.uk/article/160306/fifa_gets_onside_over_world_cup_web_pics

I am not a number – I am a tag
Synopsis

We are all associated with multiple 'numbers' such as home phone number, mobile number, national insurance number etc. However, tags are a much more natural way to identify people than numbers.[55] Tags are descriptors that individuals assign to objects. The ability to identify people by Tags, and then contact them through a web based mapping of tags with a set of numbers, will be revolutionary.

I am not a number, I am a tag

Currently, we use numbers to find, contact and identify people. The 'number' may be a phone number, a passport number or a social security number. In future, tags could perform the role which numbers currently perform. A tag is a keyword which acts like a subject or category. A keyword is used to organise web pages and objects on the Internet. Each user "tags" a web page or image using his/her own unique tag. An image or web page may have multiple tags that identify it. Web pages and images with identical tags are then linked together and users may use the tag to search for similar web pages and images.

In this section, we are discussing an extended use of the tagging concept. Specifically, the tag could be used to identify a person and to associate specific references/ events/ topics with that person. In the most simple form this would be their name, then a bio or CV. If you meet that person, then that tag could further be used to know who else they know 'friend of friend concept'. Over time, the tag concept could be extended to many other information elements associated with a person. Further, it's not just individuals, but companies which will be identified by their brands (same as their tags).

This change seems deceptively simple, but it has a profound impact because it could change the balance of power from the individual providers of telecoms services to the Web. It could also provide a better service to the customer; as we shall see in figure 26.

---

55 http://en.wikipedia.org/wiki/Tags

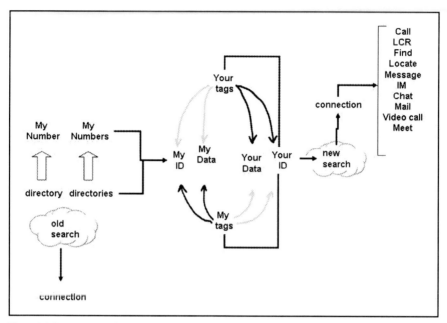

Figure 26: *I am not a number, I am a tag*

Let us look at this from a historical perspective. In the very old days, we all had just one 'number'. In fact, it was the shared household number! If people wanted to contact us, they could lookup the number using 'Directory Enquiries'. Eventually, we had more than one number. These include, mobile phone number, SkypeIn, office number, Direct dial number, home number, 'home office' number and so on. There are two obvious points to note:

Firstly, Directory Enquiries no longer has these numbers. Secondly, which of these numbers to use, often depends on where we are.

If you really wanted these numbers, you would need the person's business card (or perhaps a Plaxo card[56], LinkedIn[57] profile etc).

---

56 www.plaxo.com
57 www.linkedin.com

However, we should really ask ourselves:

- Why did we want the numbers in the first place? and

- Why do each of us have so many numbers?

The answer to the first question is: we wanted the numbers because we want to 'speak/communicate' to the person. Each person has more than one number because we have historically 'signed up' to more than one provider.

How could this process be simplified?

Here is an alternative using tags: Instead of worrying about using the Telecom operators' directory search (not knowing which Operator the person we want to contact is with), could we use a Web based search engine? Imagine, you type in a name, instead of responding with a bunch of numbers, the search engine offers you a choice of what you would like to do. It gives choices such as: Do you want to call the landline number? Do you want to send a message? Do you want to use a VoIP call? Do you want to call the mobile number? And so on...

This choice could be refined by the search engine depending on the exact location/preferences set by the person we are trying to contact. (For example, if they are mobile, the mobile number is returned first; if they are in the office, a landline number is returned; certain people do not have access to the mobile number etc).

You click yes. The search engine has now become the Operator, not by offering infrastructure but by offering the directory resolving-feature.

So, now I am no longer a 'number': I am a tag.

So why tags? Tags are merely a taxonomy. But, it is a taxonomy produced by ordinary folk - otherwise known as a "folksonomy". Tags are essential to user generated content because they allow people to create their own definitions rather than the approach being top-down. This is a level of abstraction which is more attuned to the Web. Many more relationships can be inferred from tags by the Web than from numbers.

None of this depends on knowing a number or how connection happens and it is certainly not fixed mobile convergence!

There is a role for an intermediary (the search engine), but it is transparent to the user.

The same ideas could extend to the corporate world. Corporates discovered many years ago that Vanity numbers worked. These being '0800 Flowers etc'. There was no need to remember the phone number, you could type in the name on an alpha numeric keypad. The same idea works well on mobile devices in the form of 'shortcodes'. In its ultimate form, we should be able to 'dial' from a search engine tags like COKE, BMW, TAXI or FLOWERS and get connected. In the long term, this mechanism would save on costs for reprinting different numbers and would enable the company to have the same number for different geographies. Calls could also be routed via the least cost mechanism, thereby saving the customer money.

Mobile devices will be the first to drive these changes, and will be the driver for this change. The 'tag' will be there at the point of inspiration to capture ideas, but also there when you need to connect and find, without the requirements to have it all stored locally. There are many questions yet to be answered. For the legal minds, this raises issues about copyright and trademarks. Does BT, BMW or Vodafone own their tags and who has the right to use them. There will be a fine line between use in public and private spaces. However, as the use of tags in this principal is about the user organising information for later recall, it is unlikely to be abusive or for fraud or deception.

The implementation of this topic is complex. In this section, we have only introduced this idea. In Part two of this book, we shall discuss it in greater detail.

Multilingual mobile access

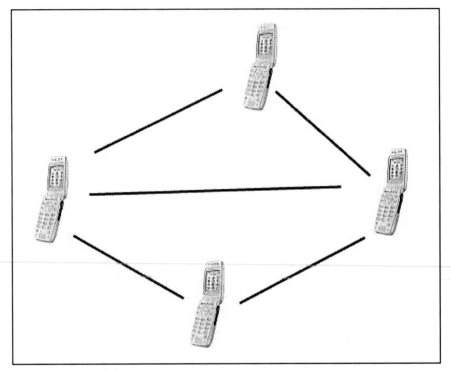

Figure 27: *A network of phones*

Synopsis

What if we could extrapolate the idea of the 'network is the computer' and extend the concepts of a 'computer' and a 'network' to higher levels in the software stack and especially to a 'Network of mobile phones'. We then have the makings of a global network which spans countries and languages. Such a network can reach many more people than 'a network of computers'.

We approach this idea in two parts:

a) The (mobile) phone network is the computer

b) The multilingual implications of this concept

The mobile phone network is the computer

Back in the 1970s, Bill Gates articulated his vision of 'a computer on every desk, and every one running Microsoft software'. By the late 90s, that vision was largely complete. Another profound vision was articulated by Sun Microsystems[58] co-founder John Gage in 1984. It concerned a different model called 'distributed computing' and it said: 'The network is the computer'. By the late 1990s, that vision was also almost realised, especially with the rise of the Internet.

Both these visions were correct in their own right (and in their own time). They are also mutually complimentary: because we can have a PC on each desk running software and we can also leverage the network for 'on demand' computing power – much like the old time sharing options. Sun's vision is more interesting because it is closer to the Internet. A recent blog by Jonathan Schwartz (Sun CEO), explains this vision in the context of 'Grid computing'[59] i.e. a collection of hardware.

Thus,

a) A 'computer' is something that can process information and

b) A computer is not confined to a single machine but to a network of connected intelligent machines.

While the notion of the 'Network is your computer' is enticing, it is still at a raw, processing capacity level.

---

58 www.sun.com
59 http://blogs.sun.com/roller/page/jonathan?entry=the_network_is_the_computer.

What if we could extrapolate the ideas of a 'computer' and a 'network' to higher levels in the stack and especially to a 'Network of mobile phones'?

While a grid network is at a physical architecture level, at higher (application) levels of the stack, another 'network' was taking shape. Called 'web services' or SOA (Service Oriented Architecture), it is a way for discrete applications to collaborate with each other, thereby creating a new, more complex application which is greater than its components. According to the Wikipedia definition[60]: *In an SOA environment, nodes on a network make resources available to other participants in the network as independent services that the participants access in a standardised way.*

With the coming of Web 2.0, lightweight versions of the SOA architecture have the capacity to reach (include) more people.

In both these cases (SOA and Web 2.0 networks), there is:

a) Some 'computation' at the application level and

b) There is a 'network' which facilitates that computation.

Purely by being simpler and at higher levels of the stack, its 'reach' is greater.

Now, if we extend the same thinking to 'Mobile Web 2.0', the 'network' becomes even more interesting (and far reaching).

a) Globally, at the end of 2005, there were 2.1 billion mobile phones vs. 1.0 billion Internet users. Even amongst those one billion Internet users, over 200 million of them accessed the Internet via a mobile phone (mostly in Japan, China and South Korea).

b) The mobile phone is capable of capturing far greater quantities of content because the phone is present at the 'point of inspiration'.

---

60 http://en.wikipedia.org/wiki/Service-Oriented_Architecture

c) Content created from the mobile phone is increasingly being 'tagged'. This makes the network of mobile phones capable of some computation/ intelligence ('Intelligence', being defined here as 'collective intelligence'/ 'wisdom of crowds').

d) Content captured from mobile phones could incorporate sound, video, pictures, podcasts and text.

e) The 'non-textual' web combined with the numbers of people using mobile phones as opposed to computers, has greater potential of being a truly global network,

f) Sound (music), images and video are much more universal than text (language). The non-textual web lends itself to being a 'network of networks'. What does this mean in practise? Think of an application similar to 'Flickr' or 'YouTube' but including images and video from all over the world. Indeed, even now, there are images and video from many different countries in an application like YouTube, but these are created mainly by travellers to those countries as opposed to residents in the countries.

Thus, in a Mobile Web 2.0 context, we could say '**The mobile phone network is the computer**'. Of course, when we say 'phone network'; we *don't mean the 'Mobile operator network'*. Rather, we mean an open, <u>Web driven,</u> application capable of aggregating (mainly non-text) content from any phone anywhere in the world. The exchange of information takes place mainly via the Web. This can be depicted as in figure 28.

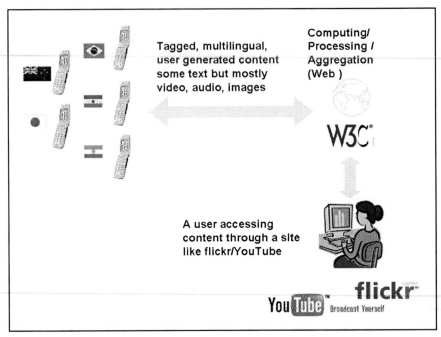

Figure 28: *Summary – The mobile phone network is the computer*

And the ideas can be summarised as per figure 29.

What if we could extrapolate the idea of the 'network is the computer' and extend the concepts of a 'computer' and a 'network' to higher levels in the software stack and especially to a 'Network of mobile phones'.

Phones	Mobile Web 2.0 / The 'phone network' is the computer
Lightweight apps	Web 2.0 / lightweight applications network is the computer
Applications	SOA/ Applications network is the computer
Grid	Network is the computer

(left vertical label: Computers) (right vertical label: Networks)

Figure 29: *Network is the computer*

Ajit Jaokar and Tony Fish

Multilingual implications

In the above discussion, the aggregation/computation is distributed. In its ultimate scale, the nodes become global and multilingual. However, the application is still driven by the Web (and not by the mobile network). In this sense, we believe that it is a Mobile Web 2.0 application, as opposed to a mobile application. This idea also has multilingual / global implications.

Multilingualism means being proficient in several languages simultaneously. Traditionally, to make an application multilingual, it had to be localised. Localisation, is the act of translating an application into a local language and adapting its content accordingly. Localisation is a complex and an expensive exercise. It is also too 'top down' and subject to changing content.

However, there is another way mobile phones can act as a global driver for unifying multilingual content. To appreciate this concept we have to understand two ideas:

a) The mobile phone is becoming more powerful and

b) The mobile phone is a "security blanket", especially for overseas travellers.

The mobile phone is fast becoming a 'security blanket'. This is more so especially if it can work overseas and *further if it can provide multilingual support overseas*. Current phones are not always capable of working overseas and lack the processing power to perform localised natural language processing[61]. However, that could change in the future. One such scenario is depicted below. A majority of the processing and information retrieval is done on the Web. However, the query is processed in the user's home language and the results are also delivered to the user in their own language.

---

61 http://en.wikipedia.org/wiki/Natural_language_processing

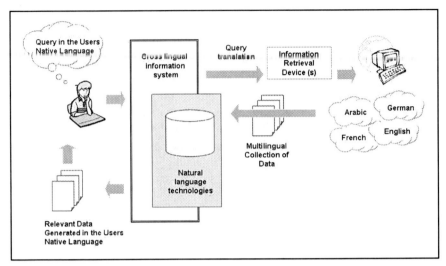

Figure 30: *Multilingual support*
Source: *Adapted from www.eurescom.de* [62]

Conclusion

A combination of the two scenarios depicted above makes the mobile phone a true integrator of people all over the world.

Digital Convergence and Mobile Web 2.0Web 2.0
Synopsis

Mobile Web 2.0 can facilitate digital convergence, potentially through mashups.

What is Digital Convergence?

Digital convergence is a much-maligned concept. Mention digital convergence, and it conjures up images of the intelligent fridge; a concept most people think they have no need for! But, digital convergence is an idea whose dawn is near. There is a lot of confusion about what exactly is meant by digital convergence. When people talk of digital convergence, they could actually mean different things:

---

62 as referenced in Multilingualism and Mobility – final report www.mulimob.com / Rudy De
    Waele (http://www.randomone.com/)

a) Co-mingled bits: The original definition of digital convergence as outlined in Nicholas Negroponte's 1995 book Being Digital.

b) Device convergence: One device to rule them all! Think the iPhone (A combination of the iPod and the mobile phone), Nokia N-gage etc.

c) Fixed to mobile convergence: A relatively new, telecoms specific area which is a part of a much broader concept called 'seamless mobility'.

d) Devices being able to speak to each other and share intelligence leading to a new service aka the 'Intelligent fridge' or home networks.

Besides these definitions, there is also the question of *"If all you have is a hammer, everything looks like a nail."* – as Mike Langberg so aptly put it in his article soon after CES[63]. By that, we mean: your tools (focus) determine your viewpoint of the world. The 'nail' in this case, is 'Digital convergence'. The 'hammer' is the viewpoint (strengths) from which each player is approaching digital convergence.

For example: (as per the article referenced above:)

For **Microsoft**, convergence is a software problem: to be solved using an upgrade of the Windows operating system (Microsoft's strength).

Intel sees convergence as a 'microprocessor problem', to be solved with a vague new branding program called ``Viiv'' [64] (A new version of 'Intel Inside'?)

Cisco sees convergence as a home networking problem, to be solved with, guess what, networking!

Yahoo and Google see convergence as an online services problem. To them, the solution lies through the Web browser, a common element in all devices.

---

63 http://www.mercurynews.com/mld/mercurynews/news/13567880.htm
64 http://www.intel.com/products/viiv/index.htm

Sony sees convergence as a consumer hardware problem, to be solved with consumer devices, new standards built around its own strengths like the Play Station.

No wonder there is confusion!

As expected, we, the authors, are also wielding a hammer (i.e. we are biased by our own experience) and hence see the 'nail' in light of the hammer. However, it's important to note that the only things common between all these definitions are:

a) Digitisation and

b) Communication

In other words, information must be digitised and it must flow freely. This leads to new services, which are greater than its parts i.e. greater than what the devices could provide on their own.

Digital convergence: the definitions

Let's first discuss the definitions above in a little more detail.

Co-mingled bits: The first definition, 'co-mingled bits', was proposed by Nicholas Negroponte in his 1995 book ' Being Digital '. Negroponte's definition of Digital convergence is:

> *"Bits co-mingle effortlessly. They start to get mixed up and can be used*
> *and re-used separately or together. The mixing of audio, video, and data*
> *is called multimedia. It sounds complicated, but it's nothing more than co-*
> *mingled bits."*

Another way to put it is: to a computer, there is no difference between a symphony, a voice call, a book, a song, a TV program, a shopping list etc as long as they are all digitised. The factors driving digital convergence/co-mingled bits include the rapid digitisation of content, greater bandwidth, increased processing power and the Internet. Digital convergence brings four (previously) distinct industry sectors in collaboration/competition with each other. Thus, we have Media/Entertainment, PC/Computing, consumer electronics and telecommunications industries all interacting more closely with each other than before. This version of digital convergence is happening all around us.

Terms like triple play or quadruple play are a part of this scenario. Triple play involves voice, broadband and mobile services and quadruple play adds digital TV to that mix. These are different from the origins of the cable industry dual play (one build to income streams). It now refers to one infrastructure with many services. Whatever name you call it, here are co-mingled bits in action! If everything has become digital, then the boundaries between the providers fade away.

Device convergence: Addresses the age old question.. 'Will we carry one general purpose device or will we carry many specialised devices?' Boundaries between devices are fading fast and devices are now capable of performing more than one function. It is unclear if customers would really want a single device. Most people have a view on this, and so do the device manufacturers. In March 2006, Microsoft confirmed that it was interested in a device combing the features of an iPod and a cellphone [65] and rumours of an iPhone launch are perpetually present[66].

Fixed to mobile convergence: Fixed to mobile convergence is a relatively new area. It has emerged because fixed line telecoms operators and mobile telecoms operators are each vying for customers in each other's traditional domains. Telecoms access networks are converging due to the emergence of new technologies. Thus, mobile network providers can provide fixed network services and vice versa. Services could also be converged. Thus, a user could access the same service from either a fixed or a mobile network. Fixed to mobile convergence could be a seen as a larger concept called 'seamless mobility' – the overall idea being that a customer should be able to 'roam' seamlessly between different network types (fixed, mobile, WiFi etc). Bodies like UMA - Unlicensed Mobile Access - are driving the standards for seamless mobility.

---

65 http://gizmodo.com/gadgets/portable-media/microsoft-confirms-ipod-competitor-in-deve-
    lopment-163685.php
66 http://biz.yahoo.com/bw/060329/20060329005526.html?.v=1

Device communications: The capacity for a range of devices to share information between each other. We discuss this definition in greater detail below.

Digital convergence: previous attempts

Note that, there are two elements common to all these definitions. Firstly, information must be digitised. Secondly, information must be capable of 'flowing freely'. The first part, digitisation, is relatively simple. It's happening all around us. However, the second part 'information flowing freely' is the real bottleneck because devices do not have a common standard to communicate with each other. For information to indeed flow freely there must be a common 'lingua franca' – a common standard. The big (and sadly predictable) battles are raging to control this communications medium.

The earliest of these can be traced to the term 'The Information Super Highway'. The phrase 'Information Superhighway' was popular with governments, politicians and people who wanted to exercise control. That's because, highways mean toll booths and choke points! A few years down the road, we know that the Information Super Highway is a road to nowhere! In spite of the failure of the Information Super Highway concept, there have been other attempts to create (and control?) a common standard.. with mixed results.

Consider the case of South Korea and Japan. In both these cases, communications technology is far more advanced. In many cases, we see convergence that we can only dream about in the west! Apart from other factors like cultural affinity to new technology, the biggest factor by far is a 'managed collaboration' – for the lack of a better word. For example - In Japan, for mobile devices, there has been a dominant player in the form of NTT DoCoMo leading to market cohesion. In South Korea, the government has actively managed standardisation with spectacular results.

While the results so obtained are commendable, they cannot (by definition) be global. That explains why Toyota can be the dominant car manufacturer but i-mode is not the world's preferred mobile platform i.e. Japan can export cars (physical goods) but not information based products which require adherence to standards (unless they can control the standard itself in some form).

The only other attempt we can think of is, Jini [67], from Sun Microsystems[68].

According to the original definition of Jini:

> *Jini is the name for a distributed computing environment, that can offer
> "network plug and play". A device or a software service can be connected
> to a network and announce its presence, and clients that wish to use such a
> service can then locate it and call it to perform tasks.*

Jini has not taken off. As per recent posts on the Jini web site, Jini may undergo
a change of direction.

The basis of a 'Lingua Franca'

So far, we have seen that:

a) Digitisation is happening all around us.

b) The communications mechanism facilitating the flow of digital content
   is unclear.

c) Top down approaches (either from governments or from corporations)
   – do not work on a global scale.

Our view is: *going up the communications stack may offer a better chance
of digital convergence*. At the lowest levels of the stack, devices communicate
via IP (Internet protocol). At the higher levels of the stack[69], the one common
element to many new devices is http[70]. The Web (by that we mean IP and
http) are the common elements to almost all new devices. Consider that the
following five devices shown are all running a browser in spite of their obvious
differences in form and functionality.

---

67 http://www.jini.org/
68 www.sun.com
69 http://en.wikipedia.org/wiki/Protocol_stack
70 http://en.wikipedia.org/wiki/Http

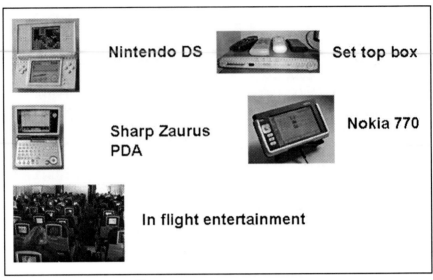

Figure 31: *Devices with browsers[71]*

Thus, the presence of a browser could offer a means to facilitate digital convergence.

Digital convergence = mashups

If we accept that many devices in future could be both IP and http capable (i.e. would also have a browser), then mashups could be used to communicate between them.

Thus, mashups could be the drivers of digital convergence.

As we have seen previously, mashups are a core element of Web 2.0. **Visual** mashups are getting all the kudos at the moment (for example HousingMaps which combines Craigslist and Google Maps), *but, mashups need not be visual…*

---

71 Images: http://www.opera.com/products/devices/markets/gallery/

Consider the Yahoo Music Engine API

Figure 32: *Yahoo music engine API*
Image source: *Yahoo[72]*

This API is used to pipe music from the home PC to the living room through the Linksys Music Bridge to wirelessly play all of your music directly to your home stereo. While the Yahoo Music engine API is relatively obscure, it could point to a future trend where device manufacturers could enable other devices to mashup with them. As hardware becomes a commodity, the ease and 'connectivity popularity' (number of mashups) could be a key differentiating factor for hardware manufacturers.

Thus, digital convergence would benefit from Mobile Web 2.0.

---

72 http://plugins.yme.music.yahoo.com/plugins/download/2005/0026/Linksys_Full.jpg

The disruptive power of Ajax and mobile widgets
Synopsis

Ajax is gaining a lot of momentum on the Web. By definition, Ajax provides a superior user interface and optimised data retrieval. However, it does a lot more on the Mobile Internet. Specifically, the disruptive potential of Ajax lies in its capability to enable the creation of Widgets. Widgets overcome the current limitations of market fragmentation and widen application distribution, making Ajax attractive to mobile application developers.

What is Ajax?

Ajax is an acronym which stands for Asynchronous communication, JavaScript, and XML. It was first proposed by Jesse James Garrett in an article in February 2005 73 Ajax is not a single technology. Neither is it a new technology. Rather, it's a combination of a number of existing technologies acting together, but in an unconventional way. These technologies are:

- XHTML and CSS for standards based presentation

- Document Object Model for dynamic display and interaction

- XML and XSLT for data interchange and manipulation

- XMLHttpRequest for asynchronous data retrieval and

- Javascript to tie everything together

(Note: we do not explain the components behind Ajax in detail. Any good book on web development or www.wikipedia.org should be able to define these).

---

73 http://www.adaptivepath.com/publications/essays/archives/000385.php

According to the Wall Street Journal, Ajax represents *"a big step toward the holy grail of having the kinds of speed and responsiveness in Web-based programs that's usually associated only with desktop software."* [74]

Indeed, momentum behind Ajax is such that it is (wrongly) equated to Web 2.0 itself. For most people, Google Maps (www.maps.google.com) is their first introduction to Ajax. The most noticeable feature of Google Maps is its superior user interface. A superior user interface has never been the forte of the Web. When compared to desktop applications, a web interface sounds primitive. Yet, to make the widest possible use of the Web, a superior (i.e. 'desktop like') interface is necessary. (As we know, this constitutes the seventh principle of Web 2.0, i.e. 'A rich user interface').

The problem itself is not new. However, in the past, it has been addressed by proprietary means and needed supplemental browser technologies like Java applets, Active-X controls and browser plug-ins. Besides being proprietary, these technologies also faced security concerns and the need to run the right version of the supplemented software.

The significance of Ajax lies in the fact that it provides a superior user interface using technologies already present in browsers i.e. it requires no supplementation.

Until Ajax came along, it was not easy to replicate the rich and responsive interaction design of desktop applications. Ajax is different from other previous attempts in addressing this problem since it is based on existing, non-proprietary standards which are already familiar to developers. In traditional web applications, most user action triggers an HTTP request. The server does some processing and returns the result back to the user. While the server is processing, the user waits! The 'start-stop-start' nature of web applications is good from a technical standpoint but not from a user interaction standpoint (since almost all user interaction is resulting in trips to the server and the user is waiting while the server is doing the work).

---

74 http://ajax.sys-con.com/read/166503.htm

Ajax solves this problem by using the Ajax engine. At the start of the session, the Ajax application loads the Ajax engine. The Ajax engine is written in JavaScript as a JavaScript library and sits in a hidden frame. The user interacts with the Ajax engine instead of the Web Server. If the user interaction does not require a trip to the server, the Ajax engine handles the interaction on its own. When the user interaction needs some data from the server, the Ajax engine makes a call asynchronously (via XML / XMLHttpRequest API) without interrupting the user's flow.

In this sense, Ajax is 'asynchronous' because the Ajax engine is communicating with the server asynchronously to the user interaction. Thus, the user gets a seamless experience (i.e. the user is not waiting). All the technologies making up Ajax are mature and stable and developers are already familiar with them. Ajax is the foundation for many new applications on the Web like Google Suggest, Google Maps, some features of Flickr and Amazon's A9.com

The classic Ajax diagram is depicted in figure 33 (source: http://www. adaptivepath.com/publications/essays/archives/000385.php)

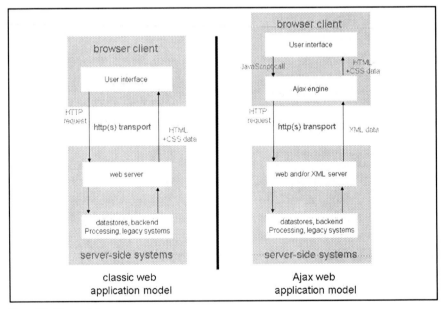

Figure 33: *Ajax model*

Ajit Jaokar and Tony Fish

The significance of Mobile Ajax

From the above discussion, we can see that - Ajax clearly solves two problems: namely a superior user interface and a standardised form of data retrieval. These two problems also apply to mobile devices and by extension, Ajax addresses them as well in a mobile environment.

However, we believe that it does far more!

Ajax is disruptive on mobile devices because:

1) **Ajax promotes the browser model of applications development**. In doing so, Ajax:

   a) Encourages the creation of 'One Web'

   b) Reduces market fragmentation

   c) Overcomes mobile application porting woes and

   d) Overcomes the issue of the walled gardens in the mobile data industry

2) **Ajax encourages the creation of Mobile Widgets**. Widgets harmonise applications development across the Web and the Mobile Web and enable a wider application distribution model. The concept of widgets is not new. It has been around for some time on the Mac. In the context of this book, a Widget is a downloadable, interactive software object that provides a single service such as a map, a news feed etc.

3) Ajax encourages the creation of **'Long tail' applications'**

4) **Ajax has the support of the developer community**.

We discuss these subjects in detail below. Firstly, we discuss some problems facing the Mobile Data Industry and then show how Ajax solves them.

Problems facing the Mobile Data Industry

MARKET FRAGMENTATION

Mobile applications are primarily consumer applications. Mobile Data Industry is an emerging industry. As with many industries in this phase of evolution, it is fragmented. To be commercially viable (especially considering the need for the network effect[75]), consumer applications need a large target audience. This can come about either by a single proprietary standard such as BREW from Qualcomm[76] (which obviously has its disadvantages) or through open standards not controlled by any one entity with few industry barriers.

To illustrate how market fragmentation affects commercial viability of a new service, we often recommend the following approach (Most of the figures needed for this approach can be easily obtained from the Web). The idea is to think in terms of 'concentric circles' in trying to find out the target audience for your application. A sample set of steps we use are as below:

a) What is the population of the country in which you are launching your application?

b) What is the percentage of handset penetration amongst this population?

c) Which operators are you targeting within this population? (Most countries have more than one Mobile operator)

d) Which handsets are you targeting within this population (not all operators support all handsets)?

e) What is the technology of deployment for example Java, SMS, WAP etc?

---

75 http://en.wikipedia.org/wiki/Network_effect
76 www.qualcomm.com

Ajit Jaokar and Tony Fish

f)   Does the application have any special technology needs such as location-based services? How many people have handsets equipped with this technology?

g)   What does a segmentation analysis of the subset reveal? (Simplest segmentation is male/female. Prepay/postpay etc)

h)   What are the channels to market for the segments we are targeting?

i)   What proportion of this subset do we expect to hit and convert to customers based on our marketing budget? (i.e. the conversion rate which can be typically 2%)

This will give you your target audience.

Hence, this target audience multiplied by the number of potential downloads per month should give you an idea of your monthly revenue. This revenue could then be tied against your cost base including your development costs, porting costs etc to arrive at a more tangible picture of success/failure of the new service. The above methodology illustrates the problem of fragmentation and it implies that very few mobile services are profitable today. Thus, we have a proliferation of 'broadcast content applications': for example, ringtones, pictures etc but very few utility applications at a mass-market level.

PORTING WOES

This problem is specific to downloaded applications (and more commonly J2ME[77] – now called Java Me). Consider the case of mobile games (a downloaded application) typically developed using J2ME.

First the good news..

- Carriers such as Sprint and Vodafone report that mobile games and other data services now account for roughly 10 percent of their annual revenues;

---

77 http://en.wikipedia.org/wiki/J2ME

- Industry consulting firm Ovum notes that there are now more than 450 million Java-enabled handsets globally, in addition to the 38 million and 15 million BREW- and Symbian-enabled handsets.

- Mobile-game publishers racked up $1.2 billion in global sales in 2004 and expect an even stronger year in 2005 as more and more consumers discover the tiny gaming consoles already in their pockets.

BUT, then the pitfalls..

Game porting generally requires developers to adapt to differences in screen resolution, processor speed, memory thresholds, and sound capabilities, all of which can vary wildly from device to device. For publishers, this can not only exponentially increase game development and asset creation time, but can also cause them to miss critical time-to-market windows in a hyper-competitive industry. This problem is so endemic that recently, fourteen of the industry's biggest hitters have joined forces to address mobile gaming's handset fragmentation nightmare by defining an open gaming architecture for native mobile games for phones. At present, one game needs to be tweaked hundreds of times to cater for a range of handsets.[78]

Effects of the porting problem are:

a) Applications development is very costly.

b) We see mainly games (i.e. few applications) in the industry at the moment.

c) 'Long tail' applications cannot be developed. We shall explain this in greater detail below.

d) Many developers, especially the smaller developers, do not have a viable business model because development costs are high and distribution channels are very expensive.

---

78 http://www.mobile-ent.biz/newsitem.php?id=782

APPLICATION DISTRIBUTION WITHOUT WALLS

The predicament of using J2ME as per the preceding example shows that it's not enough to merely set up a community process as Sun has done (which works fine as far as the technology is concerned). The technology and the applications built upon it must remain homogeneous and interoperable to enable the network effect and gain critical mass. The fewer the 'choke points' for a platform, the better it is for the industry as a whole. We will discuss this more in the next section (where we talk of how Ajax could address this problem).

The disruptive potential of Mobile Ajax

Before we consider how Ajax addresses these problems, we need to understand the significance of the browsing applications model.

THE BROWSING APPLICATIONS MODEL

There are two principal ways to categorise mobile applications – Browsing applications and Downloading applications. There are others (like Messaging applications, SIM applications and embedded applications) - but a vast majority of the applications we see today fall under downloading or browsing applications.

Browsing applications: Browsing applications are conceptually similar to browsing the Web but take into account limitations which are unique to mobility (for example - small device sizes). Similar to the Web, the service is accessed through a microbrowser which uses a URL to locate a service on a wireless web server. The client is capable of little or no processing.

Downloading applications (Smart client applications): In contrast to browsing applications, downloading applications are applications that are first downloaded and installed on the client device. The application then runs locally on the device. Unlike the browsing application, a downloaded (or smart client) application does not need to be connected to the network when it runs. Downloading applications are also called 'smart client' applications because the client (i.e. the mobile device) is capable of some processing and / or some persistent storage (caching). Currently, most Java based games are downloaded applications i.e. they are downloaded to the client, require some processing to be performed on the client and need not be always connected to the network. Enterprise mobile applications such as sales force automation are often also examples of smart client applications.

J2ME is the most common mode of developing downloading applications and WAP /XHTML (WAP/XHTML are discussed in greater detail subsequently) is the most common way of developing browsing applications.

CAN ALL APPLICATIONS BE DEVELOPED USING THE BROWSING MODEL?

Let us consider a hypothetical question: Can ALL mobile applications be implemented using browser technology?

The browser is fast becoming the universal client. It is fast replacing the older 'Client server' applications in the PC world. There is a crucial difference between browser based applications and pre-browser applications. In the pre-browser world, we need one type of program to run a specific type of application ('Word' to view word documents, 'Excel' to view spreadsheets and so on). In contrast, we can use the browser to view any type of application (i.e. one client for many applications). This makes application development much more optimal and less susceptible to software running on the client (in this case the PC).

Following on - we consider the browser and mobile applications..

With higher spec mobile devices, greater bandwidth etc let us consider the hypothetical question - Can ALL MOBILE applications be implemented using browsing technology?

After all, the browser works well on the PC as a universal client, why not on the mobile device? A corollary to this question could be:

a) When would you be forced to develop an application on a mobile device which is NOT run through a browser? And

b) Are there some fundamental differences with browsing on a mobile device vs. browsing on the Web?

Let's consider point (b) first. To understand the differences between browsing on the Web and browsing on a mobile device, we have to consider factors such as:

a) Intermittent connections: Unlike on the Web, the wireless network connection is relatively unstable and is affected by factors such as coverage (you lose the connection in a tunnel) etc.

b) Bandwidth limitations: For example, even when 3G coverage is available, the actual bandwidth is far less.

c) The need for data storage on the client: If the device has no (or little) local storage, all data has to be downloaded every time. This is not optimal given intermittent and expensive bandwidth.

d) Finally, and most importantly: A local application provides a richer user experience, especially for applications such as games.

There are other factors such as limited user input capabilities, screen sizes and so on. Some of the above factors are getting better (for example coverage blackspots are decreasing), but the overall user experience remains one of the most important factors. Thus, the answer to our hypothetical question is: No, we cannot develop all mobile applications using the browser only. However, as we shall discuss below, the architecture of browsing applications is changing and the distinctions between the browsing and downloading applications are not as clear cut as before.

THE AJAX LED RESURGENCE OF MOBILE BROWSING APPLICATIONS

There are a number of statistics pointing to a resurgence of mobile browsing.

a) Browsing is the number one (by far) mobile application that uses data, and interestingly, carrier decks only account for 50% of the traffic[79]

---

79 http://www.russellbeattie.com/notebook/1008744.html

b) In 2005, 28 percent of mobile phone owners used their phone to browse the Internet, up from 25 percent the year before.[80]

Ajax accelerates this resurgence. Thus, despite the issues mentioned above, we believe that Ajax will lead to resurgence in browsing applications and will overcome many of the problems we now face because:

- The level of abstraction shifts to the browser. Browser based applications are relatively easier to update and can be used across operators. This leads to a greater target audience for the application. Developers could have access to a critical mass of customers to make application development worthwhile. The browser also reduces fragmentation by providing a universal client (i.e. one client for all application types as opposed to a separate client for each application type).

- Ajax is making the browser experience and data management capabilities better. For example: consider surfing for downloadable ring tones. Instead of showing ringtone genres in alphabetical order like rock, 80s etc which requires lots of clicks by the user, it would be much more efficient to allow the user to go through a list of tones or to offer tones in a suggestion-panel. Tones could also be pulled into the page and played without the need to re-load, speeding up the browsing time and making the purchasing process faster and thus enhancing revenues.

- Developers are supporting Ajax on the Internet and by extension the Mobile Internet. Industry heavyweights are shifting their support to Ajax (for example IBM's support for Open Ajax[81]).

THE POWER OF MOBILE WIDGETS

In addition to enhancing the browser model, which itself has many benefits as we have seen before, Ajax also promotes Mobile Widgets.

---

80 http://news.com.com/Mobile+browsing+becoming+mainstream/2100-1039_3-6062365.html
81 www.openajax.net/

Ajax provides a mechanism to implement widgets on browsers. Mobile Widgets are important because:

a) The code base for Mobile Widgets could be shared by corresponding Web based Widgets. This means, the distribution channels for Web based Widgets increases.

b) Widgets could call other Widgets. Hence, more complex browser based applications could be developed.

c) Ajax could foster the creation of the widget authoring market similar to Apple Widgets or Opera Widgets.

It is easier to explain some of these concepts by considering the example of the Opera platform. Opera[82] is a leading vendor of browsers, both for the Web and for Mobile devices. Opera already supports Ajax in the 17 million shipped mobile browsers in 2005. This means that, with the existing Opera browsers you can use desktop Ajax services such as Gmail, Google Maps etc. This much is quite apparent and has been covered by other analysts before.

As per its definition, the Opera Platform comprises:

- Opera web browser running in full-screen mode.

- An Ajax framework for running multiple widgets/applications.

- Access to the phones native functionality through an abstraction layer.

However, there is more to the announcement than merely support for Ajax.

The significant elements are:

1) The Opera Platform is the first mobile Ajax framework. It also has a corresponding browser framework. Thus, it uses the same codebase on the browser and the mobile device. With minor configuration changes, the desktop/browser widget could also run on the mobile device Thus,

---

82 www.opera.com

---

from a developer's perspective, there is more than one way to monetise the widget (desktop, mobile and browser).

2) The Opera platform allows access to device APIs

To appreciate this concept, we have to consider the architectural drawbacks of the browser model (corollary 'a' as discussed above).

The browser model is document centric i.e. based on mark-up languages. In contrast, downloaded and native applications are application centric since they are based on a programming language.

To be really useful, any mobile applications development model must have the capacity to access data elements that are tightly coupled to the device itself. These include the telephony API, phone book, text messages, the messaging API, call records, the SIM card, the calendar, the Bluetooth stack, media player, the file system and so on. Applications running on the phone can access these services through APIs. For the most part, applications running on browsers could not access these functions (with the exception of a few proprietary solutions).

The Opera platform is a browser based programming environment that abstracts the native device APIs through a set of Javascript APIs and thus provides developers access to the low level functions on the device from the browser. This is significant and has the possibility of creating richer browser based applications. One would argue that this approach is not a purely browser based approach (because the client in this case needs some form of 'software' running locally). Note that – the platform approach is not the same as using a plug-in because a plug-in can be downloaded. In contrast, the platform is a part of the device itself and is installed by the manufacturer.

We agree that this approach is not 'pure'. However, developers will have to deal with few (perhaps less than ten) such platforms as opposed to the 'hundreds' of variants that they have to currently contend with. Nor is this approach 'open', in the sense that an application programmed on one platform will only run on that platform. Nor is it 'endorsed' by standardisation bodies, as far as we know. The whole issue is still also too early for the Mobile operators since it has yet to flow up from browser vendors to the device manufacturers and only then to the operators. However, the W3C Mobile Web initiative may also be a standards body, in this case in collaboration with OMA. Finally, the approach is still not 'rich' enough for certain types of applications such as games (see more on games below). Nevertheless, it's a huge step forward because we now have a much more uniform playing field across operators.

Knowing what we know now about Ajax, widgets etc, let us put all these things together

Many mobile applications fall under the category of the 'Long tail'. To be profitable, they should be developed quickly, deployed widely and all at low cost. Widgets make this possible. It should be possible to create a low cost widget catering for a small audience (i.e. thousands of users as opposed to millions of users) and then have these widgets run on the desktop browser as well as the mobile browser. There in lies the significance of the Ajax approach!

We see this as a generic strategy (although its leading exponent at the moment is Opera). There are many vendors who are considering the widgets strategy on the desktop including Microsoft. It is a relatively simple matter to be able to modify the code to convert to mobile widgets.

Walled gardens and OpenGardens

Now, to our favourite topic: walled gardens and open gardens. There are two types of walled gardens: firstly the extreme case like '3'[83], which blocks access outside their sites completely, and there is not much you can do about that (but we don't give that model a lot of time to survive either!). However, the other case is an operator like Vodafone, Vodafone live[84] IS a walled garden (and

---

83 www.three.co.uk
84 http://vodafone-i.co.uk/live/

we have no problems with that) BUT Vodafone itself places no restrictions to accessing the Web from the browser.

The overall approach outlined above based on Ajax could be a way to overcome the walled gardens problem because the lowest common denominator shifts to the browser (and hence it is cross operator). Indeed, current browser technology is also cross operator, but with the Ajax based approach, as outlined above, for the first time, we have the opportunity to create compelling applications and have access to device APIs from the browser through the platform. Thus, this approach increases the target audience for any given application. This means, if the browser experience is improved, we could write browser based applications for most operators, which could access a much larger user base

Ajax is accessed through the browser. There are two ways a customer can get the browser: either the browser can be pre-installed on the phone by the manufacturer or it can be installed as a separate application. Anyone can download a browser for a Smartphone. This means, all customers can potentially install their own browser and if enough people do, we have critical mass with few 'choke points' - such as specific restrictions created by Mobile operators. In other words, a means to bypass the walled garden.

How does this approach contrast with Java ME, Symbian etc

The Ajax/platform led approach closes the gap between browser based applications and downloaded applications. Thin client applications (i.e. browser based applications) will always be impacted due to the need to remain connected in a mobile environment. But, in spite of that, the user interface is improved, access to the device functions is possible and most importantly, the potential target audience increases and development cost decreases.

Ajax has better deployment models and leans more towards open source (since Ajax components are open-source). Java (MIDP) is more deployed and better specified (for now at least). Symbian/ WinMobile offers more phone integration, but is not portable and completely proprietary. Adobe Flash is another interesting alternative, but unlike the other technologies has already proved to work with Ajax on the desktop. Hence, it's an enhancer and not only a competitor to Ajax.

The (unpredictable) evolution of Ajax

Ajax's greatest strength is its support from the developer community.

Historically, popular technologies have often morphed depending on industry support. Take the humble old 'SQL' (Structured Query Language). Every vendor (such as Sybase, Informix etc) - introduced a procedural version of SQL (PLSQL in case of Oracle). SQL and PLSQL are a contradiction in terms because SQL is set based (and thus by definition not procedural) whereas PLSQL is procedural.

However, 'procedural' SQL exists because there was an industry demand (read developer support) behind it. We are witnessing the same phenomenon here with Ajax. Ajax is taking on more than it was designed for, purely because more developers are using it.

Already, as we have pointed out above, Ajax enables creation of Mobile Widgets. In its purest form, Ajax is a technology that transfers information to and from a browser in an optimal way (i.e. asynchronously). It does not have a graphical component. But, some of its best known uses **are** graphical (for example Google Maps). Thus, we expect that support for the purest form of Ajax will be almost universal (because all browsers will support the technologies that constitute the Ajax acronym). But, Ajax itself is evolving from its pure definition. It is being used to drive RIAs i.e. Rich Internet Applications and the development environment for Ajax is also evolving into a full fledged development environment in its own right. Given the above, we expect Ajax to do a lot more on the Web and also the Mobile Web.

Notes

- We believe that the Ajax/Widgets approach will be the preferred method of mobile applications development (but not the only method of applications development).

- Ajax won't solve all problems. You still need to create a service which is useful for mobile customers.

- Currently, not all users are browsing the Mobile Internet through a full browser. WAP/XHTML are still the predominant modes of browsing, but that is changing fast.

- Very few mobile operators have tried to engage with the developer community as such. Practically the only example we can think of is source O2[85].

- The plight of small developers can be illustrated from my (Ajit's) discussions with a Korean vendor when I (Ajit) spoke at iMobiCon [86]in Seoul. The vendor had finally managed to get his game listed on a UK portal. However, that was because a Korean aggregator managed to get a deal with a UK aggregator. Thus, he now had two aggregators and one Operator taking a slice of revenue! Leaving him with very little. A sad state of affairs. Surely, there must be a way to create and distribute applications globally i.e. you should be able to write for the browser and anyone who uses that browser can download and run your application.

- The browser model as outlined above provides the ability to overcome the issues which currently plague the gaming industry. Currently, we have a 'rich' interface at the expense of a revenue model. We believe that it's not necessary to have a rich interface to develop a successful mobile game. In fact, successful games can be developed using very basic technology such as SMS. Increasingly, casual games are becoming

85 www.sourceo2.com
86 www.iMobiCon.com

significant on mobile devices. Mobile gaming is different from console/ PC gaming and applying concepts from PC/console gaming to mobile gaming (such as the necessity of a rich interface provided by native applications), would be missing the point. It's far more important to foster a revenue model and / or community, both of which are possible via the browser model complemented by Ajax.

Conclusions

Ajax, Mobile Web 2.0 and Mobile Widgets reduce time to market, encourage innovation and enable a larger target market for developers. By potentially having the ability to develop for the Web and the Mobile browser at the same time, Widgets offer a better value proposition to developers. If Widgets can 'call' other Widgets, powerful applications could be developed from simple components

The biggest factor in favour of Ajax today is the support of small application developers. Developers who stand to make money from widgets. Widgets which could potentially have a large target audience. The momentum behind Ajax could lead to the emergence of a virtuous (and disruptive!) cycle.

Mobile Web 2.0 enables location-based services
Synopsis

Mobile Location-based services (LBS) have held a lot of potential. So far, attempts to launch LBS have been 'top down' i.e. large scale, walled garden initiatives by Mobile network operators. These have not been successful and will never be. Mobile Web 2.0 techniques offer an organic, user driven way to launch Mobile location-based services. This approach has a much better chance of success.

Overview

A mobile phone knows the user's location at all times. It's the only mass-market consumer device that does so. Hence, services based on the user's location (location-based services) have been the chief differentiator for mobility. Location-based services can include a range of services such as:

- Navigation (directions)

- Emergency services (911 services)

- Field services (vehicle tracking, fleet management)

- 'Find my nearest'

- Location-based mobile advertising

- Child tracking

- Tracking pets

- Finding out the location of your 'friends'

- Mobile multiplayer games etc.

From an end user perspective, a location-based service would deliver any content filtered by location: for example, a dynamic map that mirrors a user's location or a 'find my nearest' application where every fast food outlet in a one kilometre radius can be listed. By definition, any Mobile service is 'location-based', because the location of any mobile device is always known to the subscriber's mobile operator. However, in most cases, the accuracy of the location is only to the 'cell level', and the size of each cell is not uniform. It is the greater accuracy (for example, 100m) within a cell that can enable new applications not currently possible.

Also, although location is one of the main drivers/differentiators for mobility, Europe and USA have different motivations for Location-based Services.

In the USA, the motivations for LBS are statutory. The Federal Communications Commission (FCC) issued a mandate to all US wireless carriers that required them to implement Enhanced 911 (E-911) services for all wireless users. Essentially, E-911 services help to position a caller to the emergency services such as the police or the ambulance service. In its complete form, the accuracy of the location would be down to 100m. Once such accuracy is available, it can lead to other applications apart from the positioning required emergency services. In Europe, there is no statutory requirement on the carriers. Hence, the market drivers are commercial only.

The hype

LBS has been one of the most hyped up services right from the early days.

Here are some statistics/market projections. Retrospectively, they make interesting (and rather amusing) reading

March 2001: *The location-based services market will be worth about $20 billion by 2005, according to a recent report released by Ovum,[87] a market-research firm*

Oct 2003: *Overall, operators forecast average annual revenues of $12 per subscriber from location-based services in 2005 - growing to $35 by 2008 - according to an in-depth operator survey by Cambridge Positioning Systems[88](CPS).*

And more recently..

Jul. 15, 2005

Revenues from mobile location-based services will more than double in the European market in 2005, according to telecommunications industry research firm Berg Insight.[89] The firm projects a 153 percent increase in location-based services this year to 274 million Euros, and they forecast that the next five years will see an average annual growth rate of 84 percent.

June 2005, ABI research: *"When it comes to LBS, we've moved quickly from a walking pace to a run," ABI Research analyst Kenneth Hyers said in a statement.[90]*

Nor has the hype been confined to market analysts. The biggest culprits have been marketers and advertisers enticed by the possibility of sending personalised messages to users. And these are the legal ones! We can be sure that there are a whole bunch of spammers waiting in the wings. Right from the outset, advertisers have been drooling about the mythical 'Starbucks' application. (Note: As far as we know, Starbucks themselves have never tried

---

87 http://www.informationweek.com/827/ibm.htm
88 http://www.3g.co.uk/PR/Oct2003/5950.htm
89 http://www.windowsfordevices.com/news/NS6661129510.html
90 http://www.techweb.com/wire/mobile/164902024 June 2005

---

any such application. They have wisely stuck to more useful applications like WiFi access from their coffee shops).

In a nutshell, the application is supposed to work like this: As soon as you pass a café, say Starbucks, you get a mobile coupon for '10% off your next cup of coffee'. The implementation of this service is not trivial and the application itself is not cost effective because the profit from the coffee is not worth the discount. Hence, it never took off. But that did not prevent the privacy advocates[91] from crying wolf. All the hype cannot hide the fact that LBS is a very complex application. *To truly understand LBS, we have to understand both the market issues and also the technical issues. Both the marketers and the privacy advocates missed this point.*

Hence, in the following discussion, we will consider both the technical and the user aspects of LBS. We will first try and understand LBS and the issues hampering its uptake. We then discuss how Mobile Web 2.0 services could provide an alternative using an organic (user led) approach, in contrast to the existing 'top down' approach.

Features of a Location-based service
From the Mobile network operator perspective, LBS starts with determining the 'x and y' co-ordinates of the subscriber to certain accuracy. However, merely determining the 'x-y' co-ordinates is not enough. To provide a full value added service, we need three elements:

- **Position Determination** which return the 'x-y' co-ordinates, i.e. techniques to understand where the user is located

- **Location Management platforms** software that interfaces with different position determining techniques, providing independence from a specific positioning determination method and

- **Geographic Information enablers:** mechanisms to translate the location co-ordinates into information that is useful to end users, such as street addresses.

---

91 http://www.privacyrights.org/fs/fs2-wire.htm

Ajit Jaokar and Tony Fish

From a user perspective, LBS starts with the application. Thus, an end-to-end view of a location-based service includes:

- The application that the consumer will interact with from their handset.

- A means to determine the location of the subscriber's handset (positioning technologies).

- A means of managing location requests from the Mobile operator's side.

- A way of mapping that location on to content, such as a map.

- A method of delivering content back to the consumer, filtered by their location.

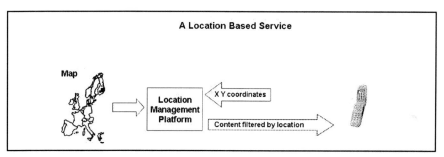

Figure 34: *A location based service*

As a first step, the Mobile operator identifies the location of the subscriber's handset. The Mobile operator is the only entity that knows the user's location (unless the Mobile operator releases that location to third parties that take location feeds from one or more Mobile operators). This location is then mapped onto a content feed that the operator may take from a number of content providers (such as a mapping company). The resultant information is sent back to the end user.

We now look at the elements of a location-based service in more detail.

POSITION DETERMINATION TECHNIQUES

Determining location comes down to knowing co-ordinates (i.e. positioning). There are broadly two types of positioning techniques: Terminal based positioning techniques and Network-based positioning techniques. Besides those, we have hybrid combinations of the above two methods. (Note that in this context; 'terminal' is likely to be the handset in most, but not all cases. For example: for 'in car' navigation systems, the terminal is a device in the car.) In general, terminal-based methods are more expensive than network-based methods but they are also more accurate. The choice of technology is based on factors like accuracy, regulatory issues (E-911), frequency of update, latency and device availability.

Terminal-based solutions:

Terminal-based solutions depend on some processing at the terminal to determine the location information. These include:

GPS: (Global Positioning System) which uses a network of satellites to locate a user's position. Although used for a few years in vehicle navigation systems, this technology is only now being used with Mobile devices. For GPS positioning to work, the terminal has to be in the line of sight of three or more satellites. Obviously, this has limitations: for example, GPS systems cannot be used indoors.

A-GPS: (Assisted Global Positioning System). Also uses the satellites as in GPS, but complements the satellite readings by using fixed GPS receivers that are placed at regular intervals.

E-OTD: (Enhanced Observed Time Difference) uses the data received from surrounding base stations. It measures the time difference taken for the data to reach the terminal. The triangulation of the time difference gives the location relative to the fixed points (i.e. base stations). E-OTD provides position information ranging from 50 to 100 metres.

Network-based solutions:
As the name suggests, network-based solutions do not need modifications to the terminal. Network-based solutions include:

Cell Identity: This technique is the simplest method and simply provides the location within a given cell. However, the problem is: cell sizes can vary. In cities, we have smaller cells, which are densely packed. In rural areas, we have fewer cells covering a large area. The accuracy of this method depends on the cell size. The smaller the cell size, the better the accuracy. Accuracy ranges from 10m (a micro cell in a building) to 500m (in a large outdoors macro cell). This is a good and a cheap solution for proximity type services (find my nearest) where high accuracy is not required. The caveat of course is, accuracy depends on the size of the cell.

Thus, in itself, the cell id has limited use as a location technique. However, the location provided within a cell can be 'enhanced' using a technique called timing advance (TA). Timing advance indicates how far the user is from the base station within a cell.

TOA: (Uplink Time of Arrival) is conceptually similar to E-OTD as discussed above and uses triangulation techniques. It works by measuring the uplink data (the data that is sent by the terminal) to three synchronised base stations within range of the terminal (triangulation). Since TOA positioning does not require any changes to the handset, it is capable of supporting even legacy handsets.

Other methods
Both Bluetooth and WiFi can be used to determine location around a certain hotspot/access point.

Cell broadcast is a Mobile technology that allows messages to be broadcast to all Mobile devices within a designated geographical area (normally a single cell). More information can be obtained at the cell broadcast forum[92]

LOCATION MANAGEMENT PLATFORMS
Location management platforms are typically carrier grade solutions from companies such as Webraska93. We do not discuss location management platforms here in detail, since they are typically installed by the Mobile

---

92 http://www.cellbroadcastforum.org/
93   www.webraska.com

---

operator and developers can treat them as a 'black box' as long as they can interface to them through the published APIs.

GEOGRAPHIC INFORMATION ENABLERS

Geographic information enablers concerned with mapping. Location-based services are often related to mapping since they often involve overlaying location information on maps. Using geographic information enablers, a developer can request a specific section of a map to a given scale.

Once an address is known, the process of **geocoding** converts the address to an x-y co-ordinate. Often the user wants to query something in relation to the current address (**find my nearest**). However, while an 'as-the-crow-flies' spatial search is relatively easy to compute, it may not be practical from the user's perspective. For example, the topography on the straight route may include a mountain or a stream. Thus, the directions have to be converted to a route by a process called **routing**, which provides directions from one geocoded point to another. Routing is not static i.e. it takes into account a range of factors, such as traffic updates. Traffic updates can change, so within a journey, the route may need to be recalculated.

Location-based services – the issues

The elements discussed above constitute a location-based service. As you will notice, this service is top down i.e. it starts with the Mobile network operator installing a complex platform and then often updating their network itself significantly. New handsets need to be supported. All this takes time. On the other hand, the users themselves may not be willing to pay a large amount of money to support LBS. As the cost to the user goes up, the user will consider other alternatives. After all, to find your nearest coffee shop, all you may have to do is to look around! Thus, the value proposition for LBS is not yet clear.

In addition, there is the issue of population of points of interest. A point of interest on a map is simply the result of a 'find my nearest' query. How can all these 'points of interest' on a map be populated? Indeed, there exist companies that are collecting this information (for example UpMyStreet[94]), but it's not

---

94 http://www.upmystreet.com/

---

cheap to map all this information on a map and keep it updated.

For the user, the impact of populating and maintaining this information on a map means more cost. In the next section, we shall discuss how Mobile Web 2.0 can overcome some of these problems.

LBS – an organic approach
Let us recap the issues hindering LBS

1)   The lack of a value proposition for operators due to the need to upgrade equipment, install new platforms and support new handsets.

2)   The lack of a value proposition for the end users due to alternate sources of information and the high cost of a location request.

3)   Populating the points of interest.

We believe that for LBS to work, an organic approach should be taken. Currently, by first trying to set up a mobile infrastructure (including both the network components and the data), we are missing out on the whole industry segment itself. The conventional 'top down' approach does not seem to be working. LBS also need the major players (often with conflicting interests) to work closely together else the market fragments.

An alternative is an organic approach, wherein the users themselves populate much of the content on the service (both from the Web and from mobile devices) and the service is driven from the Web. This approach sits well with Web 2.0 and Mobile Web 2.0. We can illustrate the organic approach to LBS using two examples:

PLAZES (POPULATING POINTS OF INTEREST)

Plazes[95] is site which allows users to maintain location and context. From the Plazes site: *Plazes is a grassroots approach to location-aware interaction,*

---

95 http://beta.plazes.com/

---

using the local network you are connected to as location reference.
Plazes allows you to share you location with the people you know and to
discover people and plazes around you.

Plazes maps a unique key created from the user's network (based on the user's router information) to a 'Plaze' that the user annotates. (A 'Plaze' is a physical place). The user names the Plaze, they can add pictures and comments. Others who are physically present at the Plaze can add information subsequently. On every Plaze, there is a box called 'Discoverer'. The person who 'discovers' the Plaze manages this box and it can be linked to their web site or blog. Additional information on that page can be edited by others.

The Plazes site maintains privacy using a pseudonym if needed. You can optionally allow the Plazes site to 'track' you using 'Trazes'. This is a default setting. Although, Trazes records past whereabouts (history), that information is not released unless the user explicitly chooses to do so. The Plazes site contains a number of features that would be useful in a live location-based service. These include:

Visibility: You can choose to remain invisible from your friends (so no one knows your current location). You can use Plazes without the launcher, meaning you are visible, but your location is unknown. All features, obviously except the location aware searches, will still be available to you.

Centralised editorial staff: There is no centralised editorial staff. Content is user driven.

Mandatory information: about the Plaze is kept to a minimum, so as to enable easy annotation.

Not spyware: The Plazes site makes great efforts to tell us that.

Data quality and control: Anyone physically present at a location can incrementally complement or alter the information for a Plaze attached to the location. Thus, the quality of data increases with the number of users and frequency of usage. The most frequented Plazes will therefore have the best quality of information, because the information is being reviewed and updated most often.

Censorship: There is very little or no censorship. The description page for a Plaze is basically a Wiki i.e. anyone can alter it.

Compliance with geographic legal issues: Plazes complies with regulations in specific countries, where the laws require that text or images be deleted.

Obsolete and historic Plazes: The following passage is well worth a read because it encounters a problem not yet endemic, but critical in the future as we all map virtual elements to a physical world: the whole question of 'virtual archaeology'.

From the Plazes site:

> *Another issue in matters of data quality is obsolete Plazes. The unique identifier for a Plaze is the network, or to be more precise the router's mac address. Thus, if you buy a new router for your Plaze or move the same router to a different Plaze, the information is either lost or incorrect. Let's say you move into a new apartment and take your router with you. Easy you say, next time you log onto Plazes you just change you address information. Right, but what about those pictures of your old house, the comments for that party you threw last year, etc? Even worse, stay in your apartment and buy a new router. All your precious information will be gone, the history of your Plaze will be unwritten, forever lost.*

> *This whole subject is a tough one. At first we thought the solution was going to be that you can just assign a plaze a new router. But this would either open all doors for a new sport called "Plaze-jacking" or an actual human would have to decide for every Plaze if the transition is legit. Not an option. So we came up with something called a "Historic Plaze". A historic Plaze is an inactive Plaze at the same location. There is two ways for a Plaze to become historic: The discoverer can mark it historic, for instance if he moves out of his flat. This way, if the router pops up the next time it is being assigned a new Plaze.*

> *If a Plaze is inactive for more than half a year and there is another Plaze with a similar description at the same location. Historic Plazes are linked to the active Plaze's description page so you can dig down into the Plaze's history. It's virtual archaeology.*

Taking social networking to the next level: Places you have visited/are going to visit can act as a common factor to bring people of shared interests together. This may work better than explicit 'friends of friends', which drives current social business networks.

DODGEBALL (TRACKING FRIENDS IN THE VICINITY)

In 2005, Google acquired a two man company (the two men being: Alex Rainert and Dennis Crowley) called Dodgeball96. Dodgeball.com is a social networking site. Unlike other social networking sites, Dodgeball is built specifically for mobile phones. Its proposition is simple: knowing where a user is, Dodgeball can tell you who or what is in the vicinity of the user.

Users can ping their friends and 'friends of friends' within 10 blocks. Like other social networking sites, a user has to explicitly indicate who their 'friends' are. There are filters and privacy measures built in the system. For example, using 'manage friends', you can indicate only some of your friends can get a certain message, as opposed to a broadcast message to all friends. Also, the location history is never revealed (so as to avoid stalkers).

Conclusions

Thus, we believe sites like Plazes and dodgeball, which rely on the user to create the metadata have a much better chance of success at LBS than the current 'Top down', Operator led initiatives. But, sadly, the Mobile Data Industry continues to persist with its bullish forecasts. We leave you with a statistic from Juniper research (June 2005)97: Mobile Location-based Services Finally Begin to Deliver: $1bn This Year to $8.5bn by 2010

---

96 www.dodgeball.com
97 http://www.lbszone.com/content/view/105/2/

Ajit Jaokar and Tony Fish

Mobile search
Synopsis

Mobile search will be a key driver for Mobile Web 2.0. But it will differ from traditional web based search.

Introduction

The vast array of information and knowledge on the Web, which by its very nature is largely unstructured and un-referenced, has given rise to the search industry.

In the public Web, the search result in itself has no intrinsic or monetary value, but rather has an applied value as a proxy for the user's 'consumption intent'. This monetisation of 'potential consumption intent' has given rise to the rapid commercial success and corresponding valuations of industry leaders such as Google, Yahoo, MSN and others. Their high stock valuations are based on the concept of exploiting the user 'intent' on an individual basis by presenting paid for marketing banners/adverts/links which are strategically placed on the returned searches. The growth of online advertising (and in the future mobile advertising) driven by the fact that the companies who are advertising are now paying on a 'per click' basis on the ad, banner or placed link. Thus, advertising expenditure is directly correlated to the response and therefore measurable.

In this chapter, we will first discuss various components and facets of search and then apply those principles to mobile searches. Our goal will be to discuss how mobile search is an important component of Mobile Web 2.0.

In a nutshell, all the search companies currently perform the following functions:

- **Crawl:** Traverse through the all the Web pages and corresponding textual (and increasingly audio and video) content on the Web and attempt to structure the information using some indexing parameters like keywords, file-types etc.

- **Index:** Index the information for retrieval, sorting and 'interpreting' the Web library through tags by genre, web links, words on pages, etc.

- **Query processor:** Process user queries to interpret the intent behind the user key word(s) search. Present the information in order of relevance to that search and intent.

Search engines differ is in their approach to the above functions; especially in how they perform indexing and in use of automated rules for run-time query processing. For instance, Google uses page ranks, where the importance of the source is determined by the number of URL references to it. Yahoo achieves the same result through tags and prompts additional user input for context. In almost all cases, search improves with previous knowledge of what the user has searched for.

However, Search is still in its infancy because it lacks the ability to accurately determine the context of intent. This causes two fundamental issues today:

a) **Lack of metadata for the source data:** The information on the Web is currently un-referenced at source. Metadata needs to be associated at source to the information entered by the user. If metadata were present at the source, the results could be presented according to the context of the query. The results could also be indexed in a certain way depending on any indexing parameters entered (for example current and historical results). For the public search, metadata fields can consist of information fields such Date, Time, Context, and Location (location being highly relevant for local searches and for the mobile web). Date and Time parameters for historic or current value (e.g. news) can be timestamped. Location, which will be required for local searches, can be interpreted/prompted through assignment of geographical/postal codes. Even a simple mechanism such as tagging can bring meaning to rich media and differentiate between similar products or information sources. How you tag a picture or audio recording could determine the context of the content. For instance: for a sunset on a beach in Goa, you could tag 'sunset', 'beach', 'holiday' or 'Goa'.

b) **Lack of context of intent:** The search parameters lack context of the intent. Context of the intent will require explicit user input. These, when matched with context aware metadata, can produce a much more relevant result.

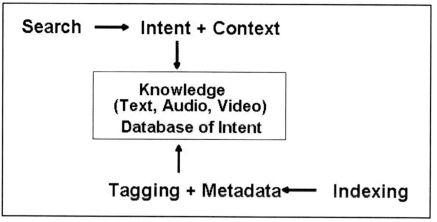

Search ⟶ Intent + Context

Knowledge
(Text, Audio, Video)
Database of Intent

Tagging + Metadata ⟵ Indexing

Figure 35: *Searching the database of knowledge*

Understanding the underlying 'intent' behind the query

User searches typically use 1 –3 keywords where the user intent is either non-specific in nature or consists of commonly used terms, popular goods, brands or items of information. Often, search is used for discovery where the users don't know exactly what they are looking for. In that case, the user is not seeking an accurate response but rather a recommendation from the search engine. Some search providers like Google use rankings by web-links (the more links that point to the website the higher the importance or relevance attached) or by 'inferred' intent from retail consumer purchase habits (history). Currently the majority (~90%) of users do not click past the first page of presented results.

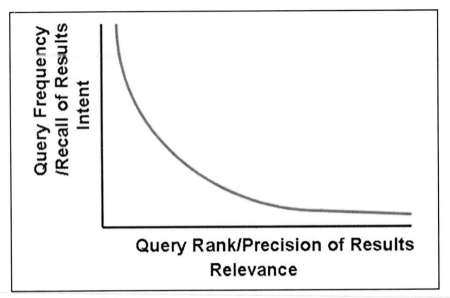

Figure 36: *The long tail of search queries*

Public search (query) keywords and the frequency of use (presented search) demonstrate a long tail effect: popular items, information and websites are most frequently searched for and presented at the top. Thus, search today can be described as generalised and populist in nature where the user's intent is 'inferred' through the search provider's existing algorithms or 'suggested' through drill down engagement. Although search is rapidly evolving today, it is not currently tuned for the mobile user.

Hence, Search in context of Mobile Web 2.0 will need to evolve further. Users 'on the move' will want to type the fewest possible keywords and will want the search results to be individualised and localised to their geographic frame of reference. Search indexes will need to incorporate the time, event and in particular location context, as well as reference rich media content input (which will be audio and visual in nature).

Indexing & Tagging of multi-media content

Information, content and knowledge are being deposited on the Web at prodigious rates in combinations of text, audio and image (video/picture). If indexing and context determination of unstructured textual based content has proved difficult; audio and video will be more so. Structured metadata will need to be appended with 'human' input and interpretation. With end user based tagging, different users will append different interpretations, emotions and contexts. The end result will be a richer set descriptors with user taxonomy or 'folksonomy'. This will allow others to arrive at the same result from different frames of intent and corresponding search queries.

One size fits all no longer applies, and so will search as one search engine or approach cannot fulfil the needs and expectations of different industries, tastes, media, localities, communities and individuals. The parameters, attributes and structure required to index a retailer or product are markedly different to those for media broadcasting - with differing user needs and behaviours. This will lead to a number of different methods of searching, as we shall see in subsequent sections.

Semantic search and Linguistics

Keyword based search is a limiting factor as use of search becomes a necessity to discover and recover information on the net. The importance of search today can be defined by the fact that, time spent online searching is on par with email. Keyword search is not intuitive and doesn't reflect or articulate the underlying intent behind the query. Next generation search will be need to be semantic in nature. It will be more like saying: 'I want to find a friendly bistro for dinner in the West End' compared to entering keywords: 'Restaurant London'. Search will also need to perform linguistics and interpret mis-spelling, especially so when one is on the move using the mobile.

Query processing will need to personalise search to the individual, perhaps by tracing the user clickstream or textstream on the mobile device to build a 'memory' for recovery of where you have been and what you have done. In turn, this will anticipate where you want to go (navigation).

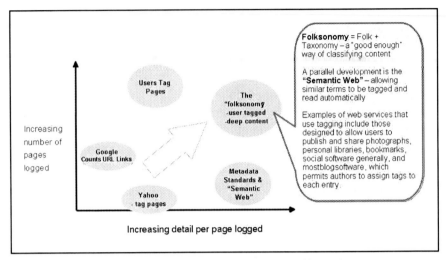

Figure 37: *Improving intent and relevance through indexing, tagging and human input*

Vertical Search

A new wave of search providers are emerging specialising in verticals, within media such as Video (Youtube) or by sector such as Retail (Amazon), Entertainment (Onmeta), News/Podcasts (Feedster). Google is also branching out in verticals. Leading media and entertainment companies are also adding search as a priority to make their content more accessible.

Social Networks

Search utilising the creative and referential power of social networks such as MySpace.com are gaining in prominence where human taxonomy (folksonomy), imitates the word of mouth in the on-line connected world. Google Co-op, Google Notebook, Jeteye, Plum, Keeboodle, Prefound, etc, allow user personal tagging, search histories, search notes to be available to other users. It is anticipated that News Corp will monetise MySpace by augmenting it with search capability.

Localised search

Localised search is the search equivalent of the yellow pages directory and newspaper classifieds, the intent mapped onto relevance of local results is highly valuable in conversion rates of consumption. Yellow pages providers,

Google Local, and Local Newspaper Owners provide localised results using postcodes filters etc to bring localised and proximity based relevance.

Peculiarities relating to Mobile Search

Now, we shall see how these principles apply to the Mobile Data Industry. Ideally, search within the context of Mobile will need to cater for the mobile user on a one-to-one basis, be personalised and be tuned to the user's lifestyle. Each search will be 'unique' in context of time, event and location within the long tail of the search curve. The transactional and monetisation value of each query in the long tail will be highly valuable because the 'intent' can be converted into call to action driven by time, event and location. For instance, if you search within the immediate vicinity of a restaurant, your search results could be presented to you filtered by your location. You could even 'impulse purchase' an item you needed by redeeming a mobile coupon.

Mobile has the capability of adding location, time and attention without the user having to input any data. Location information from the device location, time from the time stamp and attention from the amount of time a user has spent in a location (for example within a good restaurant); all serve to enhance the integrity and usefulness of the search.

Mobile search activities

To understand the peculiarities of mobile search, it is necessary to understand the flow of search activity. By doing so, we can identify those areas of activity that need special attention due to the restrictive constraints of mobility. At the same time, we can understand the opportunities mobility can bring to us. The flow of search activities are set out in the figure below. As previously discussed everything starts from an Event. These events become of interest due to some form of publicity: either by word of mouth, reputation or by advertising cash. Each of us has a threshold for interesting events, our personal filter. This filter allows us to scan a newspaper and pick out the interesting articles. It is built on our own experiences and prejudices. Search as an activity, allows us to go and find those events that we want to do something with; be that recall of an activity, learning more, supporting our existing views or for entertainment.

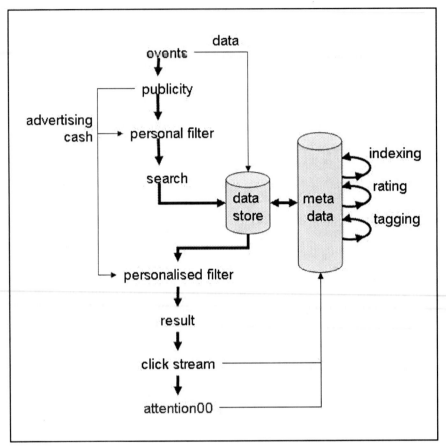

Figure 38: *Search activity*

The search tool, through the application of a unique algorithm, queries the datastore to present results. Immediately prior to that presentation, a personalised filter attempts to increase the relevancy of the result to make the experience more acceptable, timely, relevant and accurate. The results are presented. The user is then able to, in web terms, click through onto areas that are of interest. This clickstream and subsequent attention provide an ever improving feedback loop to the metadata store, that will improve the results as time goes on and the user does both more search and more clicking. An example of attention metadata is the amount of time that you take to review a web page. Metadata includes: indexing of the datastore (connecting, cataloguing and organising); rating (personal reviews) and tagging (descriptors of the datastore items).

The figure below highlights the specific areas that introduce special Mobile Web 2.0 requirements. Events, as previously discussed will be captured at the point of inspiration. This introduces a whole new data stream into the datastore in the form of private videos and images as well as voice, SMS and IM data. *The unique part will be that this data will have implicit metadata in the form of time and location*. Search will be different, as the input mechanism is not as friendly, however, this too can and will contain location data that will help improve the relevancy of the results.

The personalised filters will be now be modified to allow different styles of results to be presented depending on the device. The results will be affected as there are the limitations of the devices themselves, but the accuracy and urgency of mobile search means that the user will not be willing to spend time refining the result, as there are often easier methods of finding the data, such as calling or asking someone. The clickstream will be different, and may not be a good indicator for adverting, and attention will be lower, but could be more relevant. All these factors have to be considered when discussing mobile search, with the primary drivers for decision making in algorithms being accuracy and access method.

Search on the Mobile Device

The mobile poses additional challenges. Users on the move will type in one or two keywords, the device itself will display only 2 results or ads to be presented and the corresponding messages or brand banners will have to be truncated, i.e. 30 characters instead of 30 words. On the plus side, the search provider can take advantage of the communications capability of the devices using audio or text based queries and results. Examples include 82ASK98 and 4INFO99 which provide SMS based query and results for weather in cities, sport score, stock prices. Shazam.com performs audio search of music snippets recorded and submitted by users and returns information about the track for

---

98 www.82ask.com
99 http://www.4info.net/

---

subsequent purchase. The Mobile Marketing Association100 has established the Mobile Search Working Group to define guidelines and monetise search on the mobile. This is still an emerging space with much potential.

Conclusion

Mobile Web 2.0 search will evolve into many specialised offerings: by genre, vertical sector, social network, localised community, and media type. It will make use of attention and linguistics to personalise and understand the 'intent'. The current keyword search giants may evolve into the 'search of search' aggregators.

Mobile Web 2.0 – A service blueprint
Overview

This section gives an example of a possible Mobile Web 2.0 service. It draws from some of the work done by Dr Marc Davis and his team at the Garage Cinema Research at the University of California (Berkeley)101.

The service could be described as a 'mobile' version of a combination of del. icio.us and Flickr. Del.icio.us[102] is a site which allows you to tag (as in, add a web link you like to a central system which ranks it depending on how many other people have tagged it) and share your favourite web links. Flickr[103] is an online photo management and sharing application on the web. It allows you to upload, tag and share your digital photos.

Both del.icio.us and Flickr use tags extensively. However, note that in a mobile context, a 'tag' would have a different meaning to the term on the Web. People do not like to enter a lot of information on a mobile device.

Thus, a tag in a mobile sense, would be explicit information entered by the user (i.e. a 'web' tag) but more importantly information captured implicitly when the image was captured (for example the user's geographical location, thus producing a "geo-tag").

---

100 www.mmaglobal.com
101 http://garage.sims.berkeley.edu/
102 http://del.icio.us
103 http://www.flickr.com

Ajit Jaokar and Tony Fish

The service

The service would enable you to:

a) Search related images and get more information about a 'camera phone image' using historical analysis of metadata (including tags) from other users. This component works like del.icio.us i.e. searching via tags BUT with a mobile element because the 'tag' could include many data elements that are unique to mobility (such as location).

b) 'Share' your images with others (either nominated friends or the general public similar to Flickr but as a mobile service).

From a user perspective, the user would be able to:

a) Capture an image using a camera phone along with metadata related to that image.

b) Gain more information about that image from an analysis of historical data (either a missing element in the image or identifying the image itself).

c) Search related images based on tags.

d) Share her image with others – either nominated friends or the general public.

Let's break down the components further. We need:

a) A mobile 'tagging' system at the point of image capture.

b) A server side processing component which receives data elements from each user. It then adds insights based on historical analysis from data gleaned from other users.

c) An ability to deliver the results to the user (these could be a list of related images based on the tag or 'missing' information about the image).

d) A means to capture the user's feedback to the results.

e) A means to share images with others.

Tagging an image

It's not easy to 'tag' a mobile image at the point of capturing it. In fact, in a mobile context, implicit tagging is more important than explicit tagging. (An explicit tag being a tag which the user enters themselves).

At the point the image is taken from a camera phone, there are three classes of data elements we could potentially capture:

a) Temporal for example the time that the image was captured.

b) Spatial – The GPS location or cell id.

c) Personal/Social - Username (and other personal profile information which the user chooses to share), presence, any tags that the user has entered, other people in the vicinity (perhaps identified by Bluetooth), other places of interest recently visited etc.

The client component captures all the data elements and sends them to the server. It also displays the results from the server.

Server side processing

The server aggregates metadata from all users and applies some algorithms to the data. The data could also be 'enriched' by data sources such as land registry data, mapping data etc. It then sends the results back to the user who can browse the results.

Finding 'missing elements' of your image

In many cases, it's not easy to identify elements of the image (or in some cases, the image itself). Consider three images of the London landmark 'Big Ben' as shown below. The first image 104could include multiple tags since it contains much more information – for instance, the tags could be 'Big Ben', 'River

---

104 http://www.visitlondon.com/city_guide/river/westminster.html

---

Ajit Jaokar and Tony Fish

Thames' and 'London Eye' ('London Eye' is the Ferris wheel like structure you see in the background). The second image105 contains 'Big Ben' and the 'Houses of Parliament' (the building in the foreground). And finally, the third image 106contains only the 'Big Ben' clock tower.

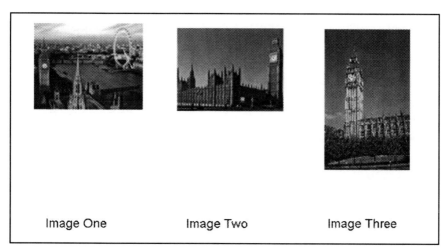

Image One   Image Two   Image Three

Figure 39: *Finding the missing element in an image*

If the person capturing the third image wanted to query 'other landmarks around Big Ben', he could get a response which included 'The River Thames', 'The Houses of Parliament' and the 'London Eye'. Note that the response is derived, based upon tags from other users (which also include Big Ben). The example is laughably trivial, but it illustrates the point! The same principles could be applied to locations which are not famous. In that case, the application becomes much more useful and richer depending on the number of users sharing information.

Sharing your images
This is the 'Flickr' component. However, 'sharing' in a mobile context, also includes location. This is very similar to an 'air graffiti' system. The air graffiti system is the ability to 'pin' digital 'post it notes' at any physical point. Suppose you were at a holiday destination and you took a picture or a video of that

---

105 http://www.visitlondon.com/whats_on/big_ben.html
106 http://www.parliament.uk/directories/hcio/clock_tower.cfm

location. You then 'posted' that note digitally with your comments and made it accessible to your 'friends'. Many years later, one of your friends happened to come to that same place and as she walked to the venue, a message would pop up on her device with your notes, picture and comments.

Like Flickr, 'friends' may be members of the general public with similar interests or a closed group.

So, is this a Mobile Web 2.0 service?
Let's consider some of the principles here:

- It's a service and not packaged software.

- It's scaleable.

- It utilises the 'long tail'; i.e. input from many users in lots of niches as opposed to a core few, which is accessible via a central server.

- The service is managing a data source (it's not just software).

- The data source gets richer as more people use the service.

- Users are trusted as 'co-developers' i.e. users contribute significantly.

- The service clearly harnesses 'collective intelligence' and by definition is 'above the level of a single device'.

- Implicit user defaults are captured.

- Data is 'some rights reserved' – people are sharing their images with others.

The two aspects not covered above are:

- A rich user experience and

- A lightweight programming model

Ajit Jaokar and Tony Fish

Further, to extend the definition to Mobile Web 2.0:

- Information is captured at the 'point of inspiration'

- Information is tagged

- The actual processing takes place on the Web

So, indeed we could call this a Mobile Web 2.0 service!

Notes

a) The example may sound trivial since Big Ben is a well known location; but the same principle could apply to images of other lesser known sites.

b) Of course, other types of data could be captured from the mobile phone for example video and sound.

c) There are no major technical bottlenecks as far as we can see (there are some commercial/privacy issues though).

d) From the above, you can see that mobile blogging, in itself, is not an example of a Web 2.0 service.

e) References:

Garage cinema research

Mobile Media Metadata for Mobile Imaging: Marc Davis University of California at Berkeley and Risto Sarvas Helsinki Institute for Information Technology

From Context to Content: Leveraging Context to Infer Media Metadata Marc Davis, Simon King, Nathan Good, and Risto Sarvas

University of California at Berkeley

PART TWO

Factors Affecting
Mobile Web 2.0

Introduction

As discussed in previous sections Mobile Web 2.0 is more than extending the 'Web' on to mobile devices. Besides the previously discussed principles, there are a number of industry trends and factors that impact Mobile Web 2.0. Therefore in this section, we will focus on these factors and discuss their possible impact on the implementation of Mobile Web 2.0.

The chapters in this section are 'self contained' and each chapter can therefore be read on its own without needing to refer to previous sections of this book in most cases. The content is based on our own experience and understanding of the trends shaping the industry. They give our views on a range of topics from Mobile Payments, Mobile TV and video, partnerships with Mobile Operators etc.

This section has been divided into the categories: *Perspective, Partnership, demographics, Devices, Services, Standardisation, Applications, IMS and Principles*. Overall, this section is designed to be read at random but we have tried to group similar categories together. The chapters are of varying lengths: some like Podcasting are quite detailed. Others are much smaller. The chapter lengths depend on the emphasis we place on that subject in relation to Mobile Web 2.0. For example: Devices are an important topic in any Mobile application. However, from our perspective(Mobile Web 2.0), a device is interesting from the functionality of capturing intelligence. This means, we are not really interested in the latest models, ergonomics etc; but rather on how devices can be used to capture content and deploy it back to the Web. Hence, the devices chapter is relatively shorter. Similarly, m-commerce is a relatively long chapter because we believe that m-commerce is a 'barometer application' i.e. an application which indicates maturity of mobile services in a country.

If you are reading this section linearly, you are likely to occasionally find 'jumps' from one topic to another. That is a by-product of the free form structure of this section. Finally, it was difficult to categorise some of these sections; for example: 'Is Mobile TV a technology or a service?' Our solution has been to call it an 'application'. The same applies to other applications (for example 'm-commerce').

Finally, more so in this chapter than elsewhere, our blogging style is apparent. We have designed these sections to be 'bloggable' and they are written in the same, concise style in many cases.

Perspective: Mobile browsing applications: In perspective

A historical perspective of mobile browsing applications with the legacy of WAP

Mobile browsing is making a resurgence especially with the advent of full browsing clients. In a purist definition, Mobile Web 2.0 is dependent on Mobile browsing. However, the legacy of mobile browsing applications cannot be ignored. Most mobile devices today do not run a full browser client. Also, in the past, browsing itself had been hyped up and this has led to some disappointing user experiences. As we have discussed before, a purist definition of Mobile Web 2.0 would comprise only of mobile browsing applications. In practise however, the Mobile Data Industry comprises of other, non browsing applications types. These include:

a) Downloading applications

b) Messaging applications SMS/MMS etc

c) Embedded applications and

d) SIM card applications

Ajit Jaokar and Tony Fish

Downloading applications (Smart client applications)

In contrast to browsing applications, downloading applications are applications that are first downloaded and installed on the client device. The application then runs locally on the device. Unlike the browsing application, a downloaded (or smart client) application does not need to be connected to the network when it runs. Downloading applications are also called 'smart client' applications because the client (i.e. the mobile device) is capable of some processing and / or some persistent storage (caching). Currently, Java based games are some of the best examples of downloaded applications i.e. they are downloaded to the client, require some processing to be performed on the client and need not be always connected to the network. Enterprise Mobile applications such as sales force automation are often also examples of smart client applications.

Messaging applications:

Messaging applications can be classed into two groups:

a) SMS (Short messaging service) and MMS (Multimedia messaging service), which originated within the Telecoms infrastructure and

b) IM (Instant Messaging), which has its roots in the Internet world.

We can view SMS and MMS as sending text or multimedia messages between two entities in a 'store and forward' mode. In other words, SMS and MMS are non real time (although they often appear to be 'real time'). IM on the other hand is a presence based, synchronous messaging system operating in near real time i.e. a connection is first formed between the two communicators and messages are sent within that session. Thus, the two communicating entities have to be physically present at the time of communication (synchronous).

Embedded applications:

Applications that are capable of running natively on a Mobile device operating system (such as Symbian applications107).

---

107 www.symbian.com

SIM card based applications:

Applications that are capable of running on SIM (Subscriber Identity Module) cards.

We shall not discuss these application models in greater detail but they are included here to give a sense of perspective i.e. to indicate that there are options to the browsing Mobile applications. Instead, we shall discuss the historical evolution of the browsing applications model. The historical evolution of mobile browsing affects Mobile Web 2.0 because, as we shall see below, browsing has had a chequered history.

The history of mobile browsing applications

Why is mobile browsing significant? Mobile subscribers outnumber the Internet subscribers 2 to 1 with an estimated three billion mobile subscribers globally by 2008. It would be useful to extend the same Internet/World Wide Web protocols on to the Mobile Internet. While it would be ideal to use HTTP/ HTML protocols over the Mobile Internet, it is not optimal to do so since they do not take into account mobility related factors such as small device size and unreliable connections over the air interface etc.

In addition, mobile browsing itself has three special characteristics when compared to traditional browsing.

a) Mobile browsing is goal based i.e. the user always seeks to achieve a goal when they are browsing. The user is not browsing randomly.

b) The user has little patience with errors and mistakes.

c) Mobile browsing is based on context. This can be explained in terms of a search on a mobile device vs. the desktop. For instance, if you searched for 'movies' on a desktop, you should get a number of responses, which include reviews, actor profiles etc.

Mobile browsing has had a chequered history, as we see from figure 40.

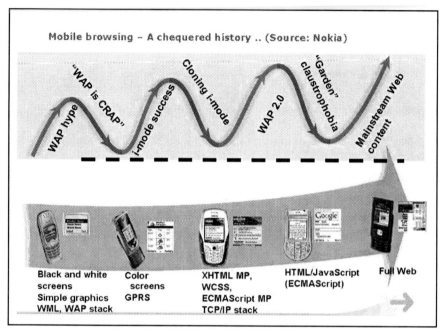

Figure 40: *Mobile browsing - A chequered history*
Source: *Nokia/Franklin Davis*

To start with, we can view the Mobile Internet as extending the conventional (landline) Internet over the air interface (also called RF network for Radio frequency network). The earliest mobile devices were 'black and white' but it was still possible to browse on these devices using WAP (Wireless Access Protocol).

The WAP forum was responsible for governing the much-maligned WAP protocol. Much has been said about WAP, most of it not flattering. However, if we see the work of the WAP forum in two parts i.e. WAP as a user interface and WAP as a transport mechanism, then the picture changes significantly. WAP as a user interface (i.e. as a browsing mechanism) never really took off mainly due to hyped up marketing leading to misguided customer expectations. However, WAP as a transport mechanism is an outstanding success and is now an integral part of the Mobile Internet.

While WAP was initially promoted as a front end development platform, it has gained more success as a transport mechanism than a language to develop browsing applications. However, as Mobile applications evolved and more Mobile applications continued to be developed. It was felt that a common browsing language was needed. According to W3C and other standards bodies, XHTML was the language of choice for developing browsing Mobile applications. Hence, WAP as a browsing mechanism evolved to XHTML.

XHTML is based on XML (Extensible Mark-up Language). Both XML and XHTML are well known standards on the Web and hence we will not discuss them in detail here. Instead, we will focus on elements that are unique to mobility. In a nutshell, HTML is a way to present data. XML is a way to describe data. In simplistic terms, XHTML is a combination of the functions from XML and HTML. The pedigree of XML is even more relevant in the wireless applications space because the foundation of XML means that XHTML is syntactically well formed. This is useful in wireless applications because well formed languages means the client (browser) has to do less work – which is ideal because the mobile client has limited resources.

XHTML-basic is a version of XHTML aimed at small information devices such as Mobile phones. More information about XHTML basic can be found at www.w3.org. From WAP 2.0 onwards, the mark-up language is a profile called XHTML-MP (XHTML – mobile profile). XHTML-MP is a superset of the W3C mandated XHTML-Basic. Inspite of XHTML providing a better browsing environment, browsing never really took off: much in contrast to Japan with its success of i-mode.

The key difference now is in the arrival of the full mobile browser from the W3C. This signals a new era in mobile browsing, which is a key driver to Mobile Web 2.0.

Ajit Jaokar and Tony Fish

Partnership: Partnerships, revenue share and MVNOs

A discussion about partnerships and revenue shares

Introduction

The Mobile data value chain lends itself to partnerships: in the sense that the final offering that the customer sees, is often created by combining information from multiple sources/companies. Thus, to be successful, companies and entities within the value chain need to understand the significance of partnerships and also the need to master the techniques of forging commercially viable partnerships.

Historically, the industry has had a good record when it comes to partnerships on an industry wide scale. For instance, GSM [108] has been a great success worldwide. On a smaller scale, each company in the value chain is partnering with others within its own value chain. As discussed previously, with the arrival of Web 2.0 and Mobile Web 2.0 coupled with the potential of mashups, the industry as a whole, is facing new challenges in its understanding of partnerships. In particular, within the Mobile Data Industry, one business partnership needs greater discussion. That is the partnership with the Mobile Network Operator. This chapter looks at the broader issues and the next chapter discusses partnerships with Mobile Network Operators in greater detail. Within this chapter, we shall first broadly consider partnership issues in general. Later in this chapter, we shall review, in greater depth, revenue shares with Mobile Network Operators and the impact this will have on the industry as a whole

Understanding partnership

The term 'partner' is used loosely in many business relationships. In its purest sense, a partner is someone who shares both the rewards and the risks. The best example we can think of is the 'Wintel partnership' i.e. Windows and Intel. In this partnership, the partners share both the upside and the downside. Both have clear demarcations, in this case Intel for hardware and Microsoft for

---

108 http://www.gsmworld.com/index.shtml

---

software. The relationship is 'trusted' to an extent that beta versions of code are shared, future products are planned in consultation and so on. Ideally, the developer wants to be a 'partner' but could end up becoming a 'supplier'. A supplier is not a partner because a supplier does not share risks.

For any business to be successful there must be a close business/supplier relationship. Strong examples of this are shown in businesses like automotive manufacturing and retail. For automotive manufacturing, parts are obtained close to production runs with the 'just in time concept'. In retail, inventory is acquired as late as possible from the supplier so stock holding and storage costs can be reduced. Nevertheless, suppliers do not share risks and hence are not in the same league as a partner when it comes to sharing the rewards. <u>Suppliers can be changed easily, partners less so.</u> Even by this token, most developers are suppliers and not partners

In the Mobile Data Industry, the question of 'who to partner with' is crucial. Developing the mobile application is just the first step, but it's much more difficult to make it into a commercial service unless you align with the right partners. Not only should partners be (ideally) channels to market but also their interests must be in tune with yours. We will explain this in more detail in subsequent chapters; especially in the context of the Mobile Network Operator.

Some questions to consider when deciding potential partners are:

- What skills do we bring to the partnership?

- Is the partner's sales cycle compatible with our own?

- What percentage of the market does the partner cover?

- What are the branding arrangements? Is your brand visible to the end customer or does the partner's brand dominate the whole service?

- What are the billing arrangements?

- What are the revenue share arrangements?

Ajit Jaokar and Tony Fish

- How is your application positioned in terms of the overall product offering to the end customer? For example: your application may just be a means to showcase the partner's infrastructure.

- Is your content/brand creating the demand or is the demand being driven by the partner's marketing/portal?

- How unique is the content/application?

- What is the level of capital investment if any required from you and the partner?

For many developers, the first instinct is to deploy the application through the Mobile Network Operators. Thus, they view Mobile Operators as channels to market. However, the real question is, ***Do Mobile Operators view themselves as channels to market for small developers?***

Later on in subsequent chapters, we discuss the long and complex answer to that question. But the short answer to the question is : *"Do Mobile Operators view themselves as channels to market for small developers?"* - is No, not really! In general, for small developers, partnerships with larger players such as the Mobile Network Operators have been VERY difficult. In fact, the difficulty in forming viable partnerships is at the heart of our advice to adopt a more web driven, browser based approach (i.e. Mobile Web 2.0). In any case, we should emphasise that there are other means to distribute a mobile application even within the Mobile Data Industry (i.e. independent of Mobile Network Operator). However, for now, let us come back to the Mobile Network Operator.

Most developers approach Mobile Operators through their developer programs or their mobile portals. Developer programs are evolving and today, we are seeing a lot more progress in this space, with many operators having revamped their developer programs. From the early days, the Mobile Operators were sending out the right signals about partnerships. But the reality was different. Certainly, the cards were (and still are) stacked largely against the small developer in the current model.

In the early days, some Mobile Operators had developer forums. In principle, they were avenues for the small developer to gain access to the Mobile operator. In practice, they were often euphemisms for selling the Mobile operator's services (for example, training and certification). Due to financial considerations, most Mobile Operators later gave up trying to seriously engage with the small developer, choosing instead to focus on safer branded content or to work with a few large developers: A risk averse strategy.

In general, the industry hasn't been exactly kind to small developers.

Under pressure from adverse market conditions, major players have changed direction 180 degrees, often forgetting the poor developer along the way. Leaving aside new companies that had developer programs and went bankrupt (like www.omnisky.com[109]) large, existing vendors are equally immune. Motorola[110], for instance, in the past, has sent conflicting signals about it's intention to support Java/Microsoft/both standards.

NTT DoCoMo and iMode

So far, the only well documented partner model to date is the NTT/Japanese model. The key elements of the NTT model from a developer perspective are:

a)   A revenue share of approximately 90/10 in favour of the developer

b)   Two classes of applications are possible; the official DoCoMo applications have a billing relationship and a set of conditions (i.e. a legal agreement with NTT). In contrast, the unofficial sites are left to their own devices but many are still profitable.

c)   NTT has significant influence and has created assembly line processes around the content creation process (la 'Toyota'). They can even influence vendors such as Sun Microsystems to influence Java standards.

While this model has been a success in Japan for a few years now, there is little evidence that it is being replicated outside of Japan. **In fact, we believe**

---

109 http://www.internetnews.com/bus-news/article.php/935801
110 http:/www.motorola.com

Ajit Jaokar and Tony Fish

that it is unlikely to be adopted in Europe and USA where the 'wholesale model' is favoured by operators. More details of this model are described in the next section.

MVNOs and the wholesale model

Developers often gaze fondly at the greener grass across the oceans to Japan and South Korea at the 90/10 model of revenue sharing. Why don't we have a 90/10 share ask many?

However, a characteristic frequently ignored about the Japanese model is the fact that the Mobile operator is not sharing revenue from carriage at favourable rates, at the wholesale level. Indeed, due to the entrenched model that already exists in Japan now, we believe that this is unlikely to happen in the future. In contrast, Europe has much better wholesale rates and a wider range of price points. Therefore, the European model is more suited for MVNOs (Mobile Virtual Network Operators) and the rise of MVNOs is already in evident in both Europe and USA.

A MVNO buys network capacity on a wholesale basis from a Mobile operator. Using that network capacity the MVNO provides services to a customer base Thus, a MVNO does not run its own network but rather 'piggybacks' on the infrastructure of an existing Mobile operator. The nature of the MVNO can range from a simple 'no frills' service to managing a part of the network function such as a dedicated Mobile switching centre/roaming etc.

MVNOs broaden the overall market base since we now have more 'channels to market' for content and applications. While the i-mode model has opened up the market in a certain way, the rise of MVNOs will create a different business environment in Europe and USA. Thus, the 90/10 arguments do not translate directly to Europe and USA because it depends on the baseline *i.e. sharing percentage of what (wholesale or retail)*? combined with what functions the MVNO takes on (marketing, customer care, billing etc). This is not to say either model is better or worse, only to observe that they are different *and potentially mutually exclusive* i.e. given one the other is not likely to arise in a specific geography.

MVNOs offer benefits from the cost side as well as from the services side by pushing new services to their existing user base. According to AT Kearney [111], MVNO margins could average between 20 to 40%. We think that net returns look slimmer at 4 to 6% but could be greater for branded portals. Thus, the wholesale/MVNO model is likely to prevail in Europe/USA as opposed to Japan where the wholesale model does not lend itself to creating a business by third parties such as MVNOs.

Hence, we see a rapid growth rate on the European MVNO scene compared to the Far East. In the UK, the partnership between Virgin[112] and T-Mobile[113] is already a success story and is drawing more entrants. The Virgin group lends the brand along with managing the customer relationship whereas T-Mobile provides the Mobile network infrastructure. But there are other reasons for the growth of MVNOs. Operators running a 'fixed only' Mobile network have an incentive to take the MVNO route to overcome the threat of 'Mobile only' vendors stealing a portion of their fixed line revenue. And some 'Mobile only' vendors can address new markets using the MVNO strategy for example the Miami based Tracfone [114]which specialises in the Hispanic market partners with 'Mobile only' operators such as Verizon[115] and Cingular [116]

There are many new entrants to this space – especially brands such as Easygroup [117] and Disney [118]. From a Mobile operator's point of view, the relationship with MVNOs brings mixed blessings and depends on the leverage the MVNO brings to the table, especially the degree of branding. Partners such as Virgin can manage marketing and customer service, saving the Mobile operator customer acquisition costs and ongoing customer care costs. By focusing on niche audiences (such as the example of Tracfone above), MVNOs enable Mobile Operators to save on the costs of formulating multiple marketing strategies for niche markets. For the MVNO, there could be some

---

111 http:// www.atkearney.com
112 http:// www.virginMobile.com
113 http://www.t-Mobile.com/
114 http://www.tracfone.com
115 www.verizon.com
116 http://www.cingular.com/
117 www.easy.com
118 www.disney.com

Ajit Jaokar and Tony Fish

disadvantages for example, not being able to manage the quality of the network. But the overall advantages outweigh the disadvantages. Retail brands, Media companies, niche market specialists like financial services companies etc are potential candidates for MVNOs.

Many innovative co-operative models are emerging where MVNOs and Mobile Operators can collaborate. One example is emerging UK based MVNO 'Subzone', which is a student based operator. Subzone aims to reduce churn at a critical point in a young person's life for example - when they leave college. Thus, the student may start at college with Subzone but as their circumstances change, they may adopt different MVNO brands but will stay with the same Mobile operator. (Because the operator will progressively market different MVNO brands at critical life changing junctures – such as leaving college). We could call this model Mobile DIVA – 'Direct and Interactive Value Acquirer'.

This collaboration works because both MVNOs and Mobile Operators are working on the same business assumptions:

1. Users spend a certain amount of money on their phone services (ARPU)

2. Users change services periodically (churn). A collaboration that increases ARPU and reduces churn is beneficial to both.

Partnership, portals and revenue shares
The issues of partners, revenue shares, Mobile Operators and walled gardens all come together with Mobile portals. The concept of portals comes from the fixed line Internet with sites such as Yahoo, AOL etc. A portal is viewed as the 'starting point'. Due to this role, any application that is at the top of the portal gets maximum exposure to the customer. Positioning on the portal is controlled by the portal owner (generally the Mobile operator). Revenues are collected by the portal owner and shared with the services on that portal.

The positioning of services and revenue share between the Mobile operator and the partners on the portal is a driving factor behind 'walled gardens'. We will discuss walled gardens in detail later. For now, we can view a walled garden as: ***an attempt by an organisation or entity to restrict or direct the*** *user in some way and control the economic model.* The controlling points are portal positioning and revenue shares. Revenue shares are negotiable and vary globally. In Europe and the USA, they depend on some of the factors we have highlighted before for example which partner is bringing more value to the table (for instance, existing brands would command a higher revenue share).

The three reference points for revenue share are: the Japanese model (with 90% to the developer), Credit card model (the Mobile operator keeps 2.5% to 4% just like a credit card company) and finally a 50/50 split. Ultimately, it all depends on building a viable business. Operators may want 50/50 or even more but if a developer cannot sustain a business at that level then everyone loses! On the other end of the scale, the existing brands want a lot more for their brand but do not realise that at these levels, the operator will find it unsustainable. This comes back to our argument for the need for considering multiple channels to market and also principles of Mobile Web 2.0 (a web driven approach).

Partnership: Channels to market
Channels to market : a wide range

We alluded before that the Mobile Network Operator is not the optimal channel to market in many cases. This leads to the question: in how many other ways can a service / application reach the customer? More than you think! It's a myth to believe that the Mobile Network Operator is the ideal channel to market for a new service/application. Here are other ways in which you can deploy your application:

- Embedded applications within the handset – for example embedded games like 'Snake' on Nokia handsets. This requires a partnership with the device manufacturer.

- Partnership programs created by operating system vendors such as Symbian. In general, many large vendors have partnership programs. Often, these programs also have a 'marketplace' i.e. a place where third party applications can be sold.

Ajit Jaokar and Tony Fish

- Third party content sites on the Internet for example Monstermob[119] .

- Distributors/aggregators who could sell games or content from their site for example www.mforma.com or in turn resell the content to operators with whom they have built commercial relationships.

- High street stores Selling applications in high street stores – either Mobile operator owned or independent stores ex Carphone warehouse.

- SIM applications i.e. applications deployed on the SIM card.

- Developer programs from handset manufactures such as forum Nokia.

- Bundling and packaging with other products - for example CD inlays for selling ringtones.

- Media / promotion / advertising companies who may manage brands. Promotion campaigns for these brands could include a Mobile element. Examples of such agencies are Flytext.

- Web portals such as Yahoo.

- Other media like newspapers, television, radio stations etc who could be potential partners due to their access to a large customer base.

- Specific channels depending on application types – for example offline dating agencies for Mobile dating applications.

- Systems integrators who could have existing relationships with large corporate customers.

- Third party Mobile portals either based on WAP or more recently Java.

- Fixed line operators (wireline, cable, DSL, WiFi).

- MVNOs (mobile virtual network operators).

---

119 www.monstermob.com

- Communities.

- Partnerships with device manufacturers (add on memory applications).

- Distribution via Bluetooth.

- Off portal revenues (Premium SMS/Bango etc).

- Prepaid cards.

- Kiosk/Point of use.

- Viral marketing and of course

- The mobile operator!!

Channels to market: A case study – mxdata

You might have a "Killer Application" (and there are many!), but you still need to sell and deliver it through the right channels. This case study (mxdata) indicates how to market an application.

Mxdata (www.mxdata.co.uk) was formed around an innovative service called 'TrafficTV'. The service consists of a Java application that could be delivered 'over the air', onto multiple handsets. The application serves real time data including video; concurrently from multiple data suppliers onto a compressed map. Although now successfully supplied through portals, websites, and internationally; the lessons we have learnt have been both painful and time consuming.

The key to our growth has been a partnership of technology, knowledge and sales all delivering success through the opening of the right channels to market. Channel choice can be as critical, perhaps even more so, than the technology decisions.

The basic channels available for a developer of mobile applications are:

- A portal

- A third Party web site e.g. Handango120 or Symbos121

- Retail

- Pre-loading

- Bespoke advertising - WebSite / Shortcode SMS

To Portal or Not to Portal?

To get a non-standard service on a portal is not straight-forward. We have seen quotes stating "£250,000 and 18 months" to get an application launched onto a portal, probably an over statement but if starting from scratch that lead time is not too far off. Even a simple game or service can be tricky to roll out and gain the exposure needed to be successful because many networks and portals are now tied into third party provisioning relationships whereby they either:

- Directly request services from 3rd parties and hence don't entertain new products

- Have outsourced the actual decision making on content to third parties.

From our experience, the portal approach has advantages and disadvantages. In summary though, we would say that unfortunately, for connected applications, the latter outweigh the former.

Disadvantages of portals

Difference in networks: All networks are different! Each works in different ways, from their commercial agreements, SLAs (service level

---

120   http://www.handango.com
121   http://www.symbos.com/

agreements), billing integration, marketing, handsets and network settings; so integrating into each portal for connected applications is often a bespoke piece of work.

Costs: Do not expect to achieve more than 50% of the net value of the service from the network. The reality is more like 25%. If they have outsourced content to a 3rd party, they will also require payment!

Billing: Currently there is a wide divergence in the capabilities of networks' billing systems to allow mechanisms such as subscription billing. There are ongoing improvements with network billing and, of course, the outsourced option is available but very expensive.

Network Settings / handsets: This can be a potential minefield. For example, recently a major handset manufacturer's handset was released with 5 different versions of firmware, all with differing Java implementations. Trying to support these firmware implementations across all networks becomes difficult and time consuming. Some networks still have the walled gardens approach, preventing any 3rd party application to be accessed, while others have multiple phone settings, each needing configuring. Developers of connected applications also need to be aware that MVNOs, such as Virgin, may not have all services of the network because they are piggy backing (i.e. don't own their own network), for example GPRS settings. The 3rd party aggregators that look after various areas of the network's portal often can be aggressive gate keepers with multiple goals, which are often at odds with that of both developer and network!

Advantages

If you do have a game, ring-tone or simple downloadable java application that can be passed over, and if margin is not an issue compared to volume, then portals do have merit. They can have a large footfall in the predominantly under 25 age group and of course, if you have a well known franchise / football team – e.g. Star Wars / Kylie Minogue/ Manchester United, then again it will work!

Ajit Jaokar and Tony Fish

3rd Party WebSites

Handango, Symbos, Motricity etc. can be cost effective, if you fit into the last category mentioned above, but again margins are low. Expect to lose between 30 to 50% of your revenue.

In addition, they are very much "boutique" styled channels; if you had The Crazy Frog or Sudoku application then people will come looking for you. However these channels are often not a good way to go for launching new services, as people will not see, or look for, services they don't know about! Another way to address this is by using the site to advertise, but again that's more cost.

Retail

We believe there will be significant growth in this area, as with PC games this will grow because, at the end of the day people do like the comfort factor of having something tangible for their money. It also enables the developer to provide a good platform for delivery of instructions!

However, retail has a large fixed cost in creating the channel but once in place there is the ability to push any number of applications as retailers look for constant sales. Margin is again an issue, with 40% + going, but high volumes and premium charging can address this.

Increasing confidence in applications and speed of download could eventually eat into this market.

Pre-loading

Pre-loading is when an application is loaded by the Mobile Network Operator on their device. This is still the "holy grail" for any developers to see a handset launch with their software on it. However, in reality it is extremely difficult, but not impossible. Our application, TrafficTV, was the first UK based application to our knowledge, to be pre-loaded onto a network's European signature devices.

Often it will be impossible as the content/marketing teams will only see a handset after it has been signed off! In addition, you have to be able to identify who is responsible for handset software because vanilla (un branded devices) sales are only running at 15%. In the rest of the cases, it's the Network who decides content, although with a vendor like Nokia, this can sometimes be reversed.

Bespoke Advertising: Websites / Shortcode SMS

If you own your own web site and are confident of your success, this has to be the most cost effective method of delivery. The caveat is; you need to build the traffic to your site. This could be achieved through partners. We partnered with a motoring organisation, a traffic provider and a map designer to launch TrafficTV – all great publicity. It takes effort and time but can prove very worthwhile. We would also recommend opening up dialogue with the media, get reviews, they can lead to cheap advertising tricks like competitions to give away your product.

The world is a big place. In this day and age there is no need to just limit yourself to one country, unless of course you have a location specific application!! Think big!

An advantage of a link with a network is that you can achieve a presence in multiple handset countries pretty quickly, but the same is the case with the international sites such as Handango.

As with all channels, knowledge is king. Understand the countries you are launching into, especially if you are in charge of the delivery mechanisms. We launched across 4 networks in America this year and found integrating 20 maps and associated live feeds of traffic about as time consuming as integrating a billing mechanism!

The Future New Technology – Google & i-mode vs. Portals

Unfortunately for the small player in this market, as in all markets, the channel is king and so finding a way to supply the masses and maintain margin will always be difficult. Such innovations as Google for Mobile and i-mode are coming along this year, but their boutique styled approach could simply lead to the big players getting more exposure, with the average developer forever living off the crumbs from the table.

So in summary, it's a tough world out there! The market for Mobile Data Applications is however a big one and growing strongly. Care just has to be taken to either "build for the channel" or invest in the knowledge and skills to develop one, remembering the lead times involved. If such initiatives are put in place, for the right products, success can and will be achieved.

Ian Tomson-Smith, Commercial director
www.mxdata.co.uk

Partnership: Partnering with Mobile Network Operators

A discussion about a special type of partnership: A partnership with the Mobile Network Operator, with the complexities and pitfalls therein

Synopsis

In the previous section, we discussed partnerships in the Mobile Data Industry. In this section, we discuss a special type of partnership: i.e. a partnership with a Mobile Network Operator. When we speak of Mobile Web 2.0 and Mobile applications development, it is not possible to ignore the role of the Mobile Network Operator. This section discusses the opportunities and pitfalls in partnering with Mobile Network Operators. To do so successfully, we need to understand the mindset and motivations of the Operators.

Thus, this chapter could be viewed as 'Inside the mind of the Operator' or seeing the industry from the perspective of a Mobile Network Operator. This viewpoint is invaluable in building partnerships and in creating new services.

In this section, we cover:

a)   The Mobile operator's viewpoint and motivation ('Inside the mind of the Mobile Operator')

b)   The interaction between mobile developers and Mobile Operators

c)   The Mobile Network Operator and the idea of 'Long tail' of Mobile applications.

Note that, this section is independent of Web 2.0/Mobile Web 2.0 and discusses the partnership between developers and Mobile Network Operators in general. You should cross-reference the information in this chapter with the sections on Channels to market and Mobile Widgets.

We start this chapter by defining some common terminology. This is important since Mobile Operators and developers use different sets of terminology, which can cause considerable confusion. Specifically, this applies to the value chain.

The value chain: A Mobile operator's perspective

Before we can understand what Mobile Operators want from independent Mobile applications developers, we need to understand how they look at the value chain.

It's important to understand that the value chain as seen by the Mobile Network Operator is very different from that seen by a Mobile applications developer because, for the Mobile operator, it includes the elements of the network itself.

The **Network operator value chain** diagram below, depicts the value chain as seen by a Mobile Network Operator. The traditional Mobile operator value chain comprises of the following elements:

- **Network equipment**
- **Network operation**
- **Terminal equipment or devices**
- **Service providers.**

As we move to a more data centric world, three more elements are added to this traditional value chain. These are:

- **Middleware**
- **Applications**
- **Content**.

Middleware is made up of two components - these are:

- **Service enabling platforms and**
- **Data middleware platforms.**

The service enabling platforms are at the <u>edge of the network</u>. Examples of such 'edge of network service enabling platforms' are: Content repurposing (taking content that was produced for distribution services such as TV or Internet and repurposing it for the size and capability restraints of the Mobile terminal). Personalisation, Real time information streaming, Voice recognition, content management and gaming applications etc. Unlike service enabling platforms, which are at the edge of the network, data middleware platforms are <u>inside the network</u>. Examples of such data middleware platforms are: Session management, Optimisation, e-wallet, Security, Messaging gateway, Location determination, Protocol conversion and OSS.

From a Mobile Network Operator's perspective, service enabling platforms and data middleware platforms form modular components. When they are combined with a software layer (application layer), they form a valuable service for the end user. A service they are willing to pay for

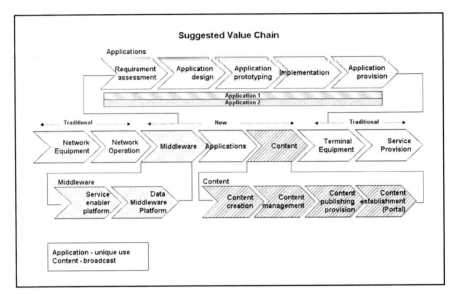

Figure 41: *Network operator value chain*

The applications layer is software that combines the middleware platforms in a unique way to produce a unique application. From the Mobile operator's perspective, this applications layer is made up of five value-added components. These components are:

- Assessment
- Design
- Prototyping
- Implementation
- Provisioning of an application

Content itself has its own value chain - made up of

- Content creation
- Content management
- Content publishing
- Context

This last part, context, could easily itself be defined as an application. Why is this so? **Because context requires knowledge about the user and the location.** An application that is able to undertake such tasks will provide the personalised context that makes it all relevant. Of course, the developer does not see this value chain, they see the more familiar value chain depicted below, which we will call the **'Delivery value chain'**.

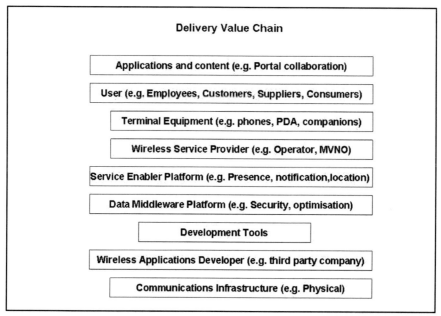

Figure 42: *Delivery value chain*

If you look at delivery value chain, figure, above, in a vertical orientation it can be described as the 'network value chain'. We start with the communications infrastructure at the base i.e. the physical stuff that makes up the network. The first block on top of this is the applications developer, closely followed by the development tools. The application developer will use tools to build an application that utilises the service enabling platforms and data middleware platform, which will be unique to the communications infrastructure.

Having built an application, the application itself is offered via a wireless service provider (which in many cases, is the same as the Mobile operator).

The user experience comprises of this application running on their terminal equipment. Or, put simply, the customer selecting this service for use on their particular handset. The reason for presenting the process in this vertical format is to show that all aspects must work and interact to produce a successful and useable application. This value chain governs the entire industry, which is depicted in the categories below.

CONTENT SERVICES

- Ringtones, wall paper, traffic, mapping, LBS, Games, Entertainment, media, news, advertising, sports, business news, video and multimedia content

INFRASTRUCTURE

- **Network components :** network components for EMS,MMS, SMS, GPS, GSM, GPRS, 3G

- **Fixed network components :** Components for Provisioning, OSS, BSS, Billing, mediation, WLL, Broadband, Bluetooth

  Devices : Handsets, PDAs, PANS, special devices, accessories

SOFTWARE AND PLATFORMS

- **Service enablement : edge of network :** Content reformatting, Personalisation, Real time information streaming, Voice recognition, Content management, Gaming, Data middleware

- **Service enablement : Inside the network :** Session management, m-payment, OTA, Security, Messaging Platforms, Location platforms, OSS

- **OS :** Symbian, Java, Pocket PC, Palm, Brew

Ajit Jaokar and Tony Fish

CHANNEL AND SERVICE PROVISION

Portals, Independent retail outlets, WASP, ISVs, System integrators, Consultants, Content aggregators

WIRELESS ENTERPRISE MARKETS

Enterprise Intranet access, Email access, Sales force automation, CRM, Inventory, Procurement

Target markets Travel and transport, Education, Financial Services, Leisure, Manufacturing, Government and Retail

The Mobile Operator's mindset

As a developer, if you wish to partner with a Mobile Network Operator, you have to understand their motivations. It is crucial to realise that the Mobile operator sees all applications in the context of the Network value chain (and not the developer value chain). As we have seen before this is different from the conventional value chain developers are used to. The Mobile operator's approach is based on maximising revenues within this value chain. Sadly, 'maximising revenue' often comes down to 'doing it themselves'. Putting it rather cynically, everything you own, Mobile Operators would like to own and exploit for no cost, everything they own, they cannot share.

Whilst this is a little harsh on the Mobile operator, it does illustrate a point. The point being, that the Mobile operator would like to do everything themselves. In practise, doing everything 'in house' is not possible. The Mobile operator seeks to balance these two processes:

Firstly, a balance between internal and external applications development.

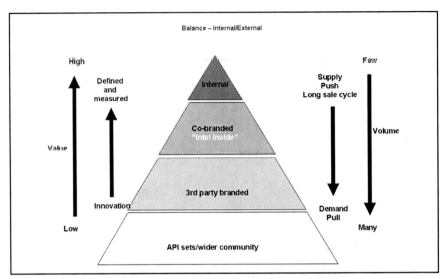

Figure 43: *Balance between internal and external*

Often, the Mobile operator's most favoured choice is to develop all applications internally by themselves. However, they have realised that they don't have a monopoly on all innovation. They have also realised that their internally developed applications tend to be late to market and are not as functional as the ones developed by a third party application developer.

Secondly, a balance between demand pull and supplier push

There is also a realisation, that innovation is generated by **demand pull** i.e. the latest new record produces a new ringtone, game, logo and wallpaper. In contrast, high value applications tend to require a high degree of market education and are **supplier pushed**, for example: e-mail on the move, integration into corporate supply chain management, field service etc.

Thus, innovation produces a very large number of low value applications. Conversely, there are only a few well-defined high value applications and these need market push. Therefore, the Mobile Operators need to find a way to balance innovation and value (i.e. demand pull and supplier push). They try to find this balance by producing a few of their own internal applications; these are augmented by co-branded and third party branded applications.

Ajit Jaokar and Tony Fish

However, the highest volume, lowest value, greatest innovations will come from a wide community of developers who produce an application in response to a specific market problem as we have seen in the case of the Long tail. This type of application will cost very little to build (relatively speaking) as it will utilise existing services and middleware. In many cases, it may not actually build anything new, but may combine existing components. Monetizing these applications presents the familiar problem of monetizing the 'Long tail'.

The problem is not new and has already been faced by the software industry.

If we look at the diagram **'Getting more value from the price/demand curve'**, we can see the corresponding situation faced by the software industry. Mobile operators also have a similar motivation and hence they seek co branded and third party applications. The important lesson for Mobile Operators and the industry as a whole is that, **partnerships can be profitable when properly managed** as we can see from the example of the software industry (Microsoft) below.

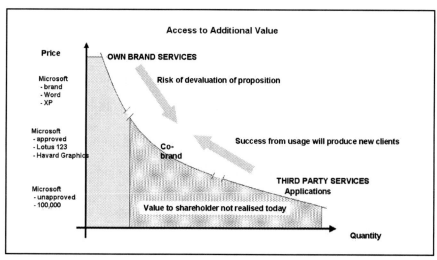

Figure 44: *Access to additional value*

For a large company, there is a lot of value to be gained by being the 'gatekeeper', even though the company (in this case the Mobile Network Operator) is not responsible for the marketing or customer service. For example, Microsoft could certainly gain value through it's own branded products like Microsoft Word or Microsoft Excel. But the ability to co-brand and deliver other applications, *can collectively create a LOT more value.*

As illustrated by the Microsoft example above, co-branded applications could include security or graphics type applications. Similarly, in a Mobile operator scenario, third party applications would involve hundreds of thousands of innovative applications that work perfectly well for the user but which may not necessarily be approved by the Mobile operator. Having seen the rationale underpinning partnerships with Mobile Network Operators, in the next section we shall discuss how developers can approach and partner with Mobile Operators.

Partnering with Mobile Operators

Developers can approach Mobile Operators in a number of ways. The 'Third party relationships' figure tries to summarise the third party relationships that the Mobile operator could have. These relationships can be classified into three broad groups:

- **Strategic relationships**: this could include corporate venturing, alliances, partnerships, research and development.

- **Supply relationships**: these are outsourcing, system integration and consulting technology supply relationships and

- **Third party developers.**

Note that: 'Third party developers' is the only category that actually increases the revenue for the Mobile operator, whereas the other two categories are components of cost.

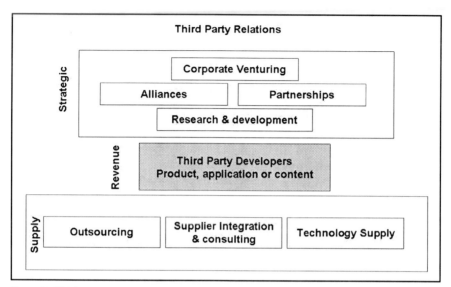

Figure 45: *Third party relationships*

Commercially, there are potentially three avenues to interface with the Mobile operator:

- **The commercial team** (headed by the commercial director)

- **The technology team** (headed by the CTO) and

- **The Mobile operator's own developer program** (if one exists).

These are usually held by different board/ executive members within the Mobile operator's organisation and these three are often in conflict. At this point, it's worth noting that developers are not the only 'Third party suppliers' to the Mobile operator. Large technology suppliers (for instance middleware vendors are also deemed to be external **suppliers. Critically, the needs and capabilities of technology suppliers are very different from those of developers.**

In addition, the various technology suppliers to the Mobile operator wish to differentiate themselves when being considered for new projects. One of the mechanisms they have for creating differentiation is to bring third party companies to their bid. Technology suppliers such as Nortel, Nokia, Motorola etc will be actively going out to source interesting applications that show their platforms or their technologies in better light. The supplier will then bundle the value of these applications or services within the sale of their own platform technology.

Typically, the CTO (technology team) will own the relationships with these technology suppliers, and the CTO sees their remit as bringing additional applications to the table. However, this will be in conflict with the commercial director who will own the strategic relationships. Most Mobile Network Operators will have a few strategic relationships, which are typically large third party companies. These relationships will be created by the Commercial director. The commercial director will see their role as 'business development' and that they should own prime responsibility for bringing additional applications to the Mobile operator that increase the value of the organisation. They do this by asking their strategic partners to find and source interesting third party application companies. Thus, it is very likely that the CTO and commercial director will source similar types of applications from different companies on different commercial terms through different partners, which will confuse the decision-making process.

Then, there is the third avenue; i.e. the Mobile operator's developer community that is also busy trying to source applications. The third party developer program, which is run by the Mobile operator itself, sees its value as sourcing third party applications. The director of the board, with responsibility for this unit, may argue against a bid from the CTO that includes additional bundled applications. This director may feel the additional bundles will confuse what the technology supplier will actually deliver within that supplier's price-packaged offer.

In the same way, this director will also be arguing with the commercial director against using their partnerships to source third party applications.

The logic behind this argument is that the Mobile operator may feel that they have a better knowledge then a third party company of the applications that customers want. Furthermore, they may feel that third party companies will only be attempting to find 'interesting' applications that are not currently demanded by customers.

Ajit Jaokar and Tony Fish

We can conclude from our third party relationship discussion that it is very unclear who actually owns third party relationships with the Mobile operator. Furthermore, it is clear that they themselves don't really know who owns the overall relationships. We could call this predicament as the ***Mobile operator's dilemma!***

The developer's dilemma

Like the Operators, the developers also have a dilemma. From the Operator's standpoint, we can see there are many potential possibilities (via many entities within the Operator's organisation) for the developer to engage with Operator. This creates considerable confusion for the Operator and the Developer. From the developer's standpoint, there is even more confusion. The developer may have to deal with many different entities to create and deliver their service. Not all of these may be within the Operator's organisation i.e. the developer is often having to deal with multiple companies when deploying their service. In other words, for the developer, there is no 'single source' of information. Nothing is available that can co-ordinate their communication with all of these different companies.

Rather like the old game the prisoner's dilemma, we have a four box quadrant that provides a difficult decision making scenario. The figure below shows us the dilemma. There is a need to work with the network equipment manufacturers, as ultimately the service will be delivered on their technological equipment. There is also a need to work with the middleware companies, as without these, their services will not be enabled. In addition, one must work with the terminal equipment manufacturers, because unless your application is designed with terminal equipment in mind, the user experience is likely to be very poor. The final component is the service provider, as without a relationship with the service provider (for example MVNO (Mobile virtual network operator) or a mobile portal), there is no mechanism for revenue collection.

This scenario assumes that the application you are developing is complex i.e. it interacts with all of the four elements discussed above.

That may not be the case (for example: your application may be a simple, single player game), however this would still mean that you need information from more than one company. In the simplest case, (single player game), you need information from the Mobile Network Operator and the terminal manufacturer (handset vendor). However, in a very complex situation (a mobile multiplayer game, to extend the same example), you need to interact with all the four layers in each network (i.e. for each Mobile Network Operator).

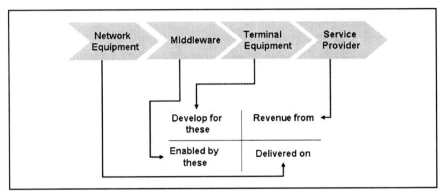

Figure 46: Developer's dilemma

There is no easy decision, hence the reason for the dilemma!

Why is this decision important? Because, without it, you can't have the network effects (i.e. the viral propagation of an application). Without the network effect, you will be dependent on huge marketing budgets to be able to build users and revenue for your application.

The developer's dilemma is one the key reasons to consider Mobile Web 2.0.

Many developers wrongly view the Mobile operator as a channel to market. We believe that the strategy of using Mobile Network Operators as channels to market is risky and can only offer limited success (since it ignores the network effects, which are difficult to achieve through Mobile Network Operators).

Therefore, we emphasise web/browser led development that offers the potential of greater critical mass.

Ajit Jaokar and Tony Fish

Points of engagement with a Mobile operator

We have seen that there are a number of points where a developer could engage with a Mobile Network Operator. But when should developers engage with a Mobile operator? Note: This assumes that the developer wants to engage with the Mobile operator in the first place. As per the previous section, for many developers , engaging with the Mobile operator may not be a viable strategy.

The **'where to engage'** figure, gives the potential points at which a Mobile operator can engage with a third party.

The **first point** of engagement is at the idea stage. The difficulty for Mobile Operators engaging with developers at the idea stage is, Mobile Operators have no skillsets /knowledgebase to pick out the stars from the duds. Indeed, if you only have an idea, taking it to a Mobile operator is not likely to produce any results at all.

The **second point** of engagement is when the application developer has produced a proof of concept. The idea now has some validation and there are some numbers, which demonstrate the magnitude of the opportunity. It is unlikely that there would be a different response at this stage from that at the idea stage, because there is still no prototype that would demonstrate the reality of the application.

The **third point** of engagement is when the prototype has been built and is ready for testing. At this point, the Mobile application developer has used development tools to produce an interactive application that can be shown to the Operator. The difficulty now for the Mobile operator is choice. By choice, we mean, which configuration of handset, platform and network will work best with the developer's application. From the experiences of existing applications developers, even at this stage, the results are not good because Mobile Operators are still unable to determine which applications will make a difference.

The **fourth point** of engagement is when the prototype has been built into a product this product has been integrated and is capable of being tested on a live network. The key issue here is that the developer has now committed a large amount cash. Given this high degree of commitment from the application developer, the application developer hopes that the Mobile operator should be able to provide some cash to cover the cost of proving and integrating their application into the Mobile operator's network

However, the Mobile operator sees it differently. They believe that the potential opportunity for the Mobile applications developer is so large that they (the developer!) should be willing to fund the cost of integrating the application with their network. (Note that the developer may need to work with more than one Mobile Network Operator. Hence, the total costs for the developer are indeed substantial).

The Mobile operator doesn't always understand the financial constraints faced by the Mobile applications developer. As the applications developer doesn't have any revenue coming in during the development process, they (the applications developers) may not have significant reserve funds to enable full development of the application. Even at this stage, the chasm between the two parties is wide and no real progress can be made.

The **fifth point** of engagement is where the developer, to overcome the problem of engagement at point four, has funded the integration and testing themselves. At this point, the Mobile developer wishes to have a contract with an underlying commitment written in. The developer justifies such commitment on the premise that they have funded, to date, all of the costs of integration, development and testing. All that remains is the marketing of the application to generate revenue. The application developer sees the Mobile operator as that marketing channel.

Yet again, the Mobile Operators don't see themselves as a marketing channel at all! Furthermore, it isn't known which application is going to be successful, so one will not be marketed in favour of another. In the past, Mobile Operators had provided commitments to smaller third party companies, which have unfortunately backfired and have cost them a significant amount of money due to under performance of the application.

Once bitten, they are very cautious about commitments. Again, we see no deal at this stage.

Ajit Jaokar and Tony Fish

The **sixth point** of engagement is where the Mobile application developer has crossed the chasm of point 4 and 5, and has a successful application that is in the market. However, the application developer discovers that several other developers have already produced a functionally similar application! The Mobile operator will wish to source the best application.

To the developer's horror, the Mobile operator now enters into an evaluation and testing phase of several competitive but similar applications! The application developer at this point must provide a differential advantage to be selected. This differential advantage could be the cost, the user experience or the lack of integration hurdles for example.

If selected at this point the Mobile application developer should enjoy a good margin. However, the Mobile operator still determines the overall margin that can be achieved. The margin that the Mobile operator often gives does not justify the overall investment required to get an applications developer to this point.

The **seventh point** of engagement, the last point, is where an application developer has an application in the market that is generating revenue. The developer can take this application to a competing Mobile operator. By having a second or third channel to market, the developer hopes to gain additional revenue.

This is the only point where the application developer has some leverage and is able to bargain from a position of strength and thereby force a better margin.

It is at this point in the development and engagement cycle where the Mobile Operators are at their weakest. The Mobile Operators have a requirement, however, the risk for the developers are also at their highest. If the developers don't get beyond this point, they are unlikely to survive. What is interesting is that the Mobile Operators perceive that there are so many Mobile application developers that they are not worried if **one or two thousand die** on route.

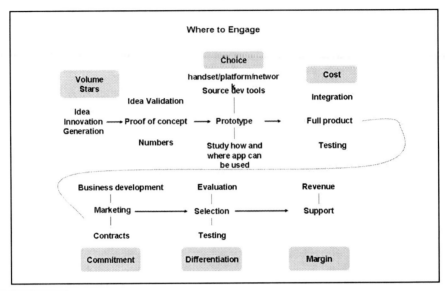

Figure 47: *Where to engage*

A lot also depends on the overall objectives of the Mobile Operators themselves.

If the objective of the Mobile operator is to provide differentiation, they can afford to engage later in the cycle. If their objective is to find innovation, they have to engage earlier in the cycle. Mobile Operators often feel that internal developments should provide differentiation and third party companies should provide innovation, as this would allow them to balance the resources needed. However, third party developers do not have a clear demarcation of where their applications fit between differentiation and innovation.

(The above paragraph is worth re reading again since it offers some crucial insights). Increasingly, the criteria that the Mobile Operators are setting imply that they only want to engage with application developers who are at the end of the cycle. Currently, we observe base level application criteria such as:

Ajit Jaokar and Tony Fish

- Must be standalone

- Will only require the network to provide carriage

- Must already have a commercial customer

- Must be of industrial grade

- Technology assessment is simple

- Has a proven business ROI

The result of these criteria means that the hurdle rate is so high that they would produce no inputs, and input in this case means a company that has developed a Mobile application. However, if the Mobile operator lowers these hurdle rates, then they may find that they are in inundated with very novel ideas, but no concrete applications that they can take to market. If they have too many novel ideas, they are unable to determine if any of them have any value. The input volume is too high and the output is zero.. To get these criteria right a careful balancing act is required.

But not all Mobile Operators are the same. Some are more dynamic, while others are risk averse. We can identify four winning characteristics in a Mobile operator. These include

- Speed of decision-making

- Focus

- Structured approach

- Sufficient resources to actually make something happen.

The converse of these could be described, as **losing characteristics.** These would be; slowness to commit, unclear position in the market, lack of focus and inconsistent behaviour. Sadly, today many of the Mobile Operators tend to exhibit the losing characteristics, whereas some of the technology providers/ system providers, who run their own mobile developer forums, tend to display winning characteristics!

Engagement scenarios

The winning/losing characteristics could be reflected in four potential engagement scenarios

- Scenario a, is where the Mobile operator is a follower

- Scenario b, is where the Mobile operator is focused

- Scenario c, is where the Mobile operator wants to be the leader in Mobile data and finally

- Scenario d. is where they do nothing.

Looking at each of these in turn:

Scenario a, is where the Mobile operator wants to be a follower, by this we mean that they would exhibit the following characteristics:

- Work only with developed applications (working, live applications)

- Turn away non working applications

- Provision of carriage only for the application: no need to do any integration

- No differentiation in service created by applications compared to others (i.e. an application of similar standing is already in the market, thereby mitigating deployment risk in the first place)

- Low risk to build applications revenues

- Integrated into a wider partner program

- Very clear external communications to manage expectations and resources

Within this area of being a follower there should be two further categorisations: that of the **active follower** and **the passive follower**. The active follower will ask their partners to bring to them new applications. In this case, the Mobile operator may choose to work with a partner on applications that don't yet meet the characteristics described above. Passive followers conversely follow tightly and rigidly the characteristics above.

Scenario b, is where the Mobile operator wants to be focused, by this we mean that they would exhibit the following characteristics:

- Willing to work with applications in the development

- Be very selective in focus

- Provide solutions

- Unlikely to have any differential offerings

- Provide high technical resources to help and assist partners in integration and delivery

- Integrated into a wider partner program

At this stage there a few Mobile operators who are displaying this characteristic, however technical players such as Blackberry would be a good example of a Mobile focused developer community.

Scenario c, is where the Mobile operator wants to be a leader, by this we mean that they would exhibit the following characteristics:

- Be willing to build an aggressive developer community

- Provide a high degree of developer support

- Will publish a high degree of API and SDK programs pre launch

- Will never second guess which ones will be successful

- Will be seen as an innovator with an innovative brand

- An example of a Mobile operator who wanted to be a leader was mmo2[122], in that they developed a program, SourceO2 [123], however it is apparent that more recently they have been moving towards being a follower of scenario a.

---

122 www.mmo2.com
123 www.sourceo2.com

Scenario d, is where the Mobile operator decides to do nothing in the area of engaging third party application developers. Their belief is that there is not sufficient value to be created in this area at this time, and that their time would be better spent developing their own applications internally rather than engaging third parties. The best example of this will be T-Mobile[124].

(Note: While T-mobile is not actively engaging with third party developers, it is very active in the MVNO space. So, this appears to be a conscious decision to focus the company in a certain direction)

Summary of engagement scenarios

The motivations and commitments required by the leader are very different from those of follower. The primary difference will be at the engagement stage. The leader will engage earlier in the cycle than the follower. They will commit tangible resources and facilities in their development programs.

A comparison of developer programs

Here are examples of some companies (Mobile Operators and system partners) who have developer programs and are interfacing with third party developers.

Mobile Network Operators
Some developer programs run by Mobile Operators are compared as follows:

SourceO2: SourceO2 [125]Provides a one-stop shop for developers to work with mmo2[126]. Source O2 is a pioneering program with events, knowledge share, test beds, platforms etc. It has 15,000 plus registered users and started off as wanting to be the leader in wireless applications. The program included developer labs, which were closed down for lack of interest. Through their revolution program, they offer distribution opportunities for applications.

---

124 www.t-mobile.com
125 www.sourceo2.com
126 www.mmo2.com

Ajit Jaokar and Tony Fish

Vodafone: Vodafone[127] via 'Via Vodafone' recently revamped Vodafone developer program provides access to the Vodafone network.

Orange: Orange developer program[128] - Another recently re invigorated program provides access to the Orange network

The system partners compared
A comparison between some system partner development programs is as below:

Blackberry: [129]Due to the large number of Blackberry devices, this program is increasingly attractive to developers. Blackberry development is done in Java, which makes the application relatively easy to develop.

Motricity: [130]The program aims to distribute applications through their Mobile operator customers. Members gain access to technical information and a platform, which enables them to get simple, downloadable applications to market without the Mobile operator.

Java One: [131]Java One enables global development of Java software similar to the Microsoft developer program. Tools and knowledge are provided to enable developers to implement and test Java applications. However, note that merely conformance to Sun's Java standards is not enough i.e. a Mobile operator may have their own tests to conform to before the application/game is accepted.

Nokia: Forum Nokia[132] provides a single stop shop to create applications for Nokia devices. It also allows access to the customer through Nokia's own distribution program. This forum is seen within the industry as a competitor to the Mobile Operators due to its own distribution program i.e. brand.

HP (Hewlett Packard) Mobile regional services bazaar: [133] The HP service

---

127 http://www.via.vodafone.com/
128 http://developers.orange.com
129 www.blackberry.com/developers
130 http://www.motricity.com/
131 http://java.sun.com/javaone/index.jsp
132 www.forum.nokia.com
133 www.hpbazaar.com

---

is implemented through a series of regional programs and partners as a place to nurture innovation. It includes support from HP's distribution, sales and marketing programs.

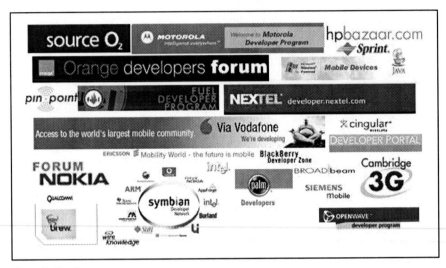

Figure 48: *Developer programs*

Operators, Mobile content and the Long tail

Our view in regards to engagement with the Mobile Network Operator can be summarised as below:

1) Engagement with Mobile Operators is a complex and risky operation.

2) By definition, you cannot leverage the power of the Network effect if you work with individual Mobile Operators.

3) How (which department) of the operator you engage with matters

4) When (at what point in your development cycle) you engage in, also matters

5) Not all Mobile Operators are the same

We believe that a web driven, Operator agnostic, browser based approach would mitigate many of these factors. This forms the basis of Mobile Web 2.0, as we have seen before.

Before we conclude this chapter, a final point needs to be considered i.e. the impact of the 'Long tail'. Mobile operator revenue is dominated by voice and will continue to be so. Voice, text, e-mail etc are homogenous services used by a majority of the 2.5 billion, and growing, Mobile/ Cellular subscribers. These services lend themselves to mass market provisioning and marketing, which Mobile Operators are familiar with. In contrast, most Mobile data applications sit within the 'Long tail' as shown below

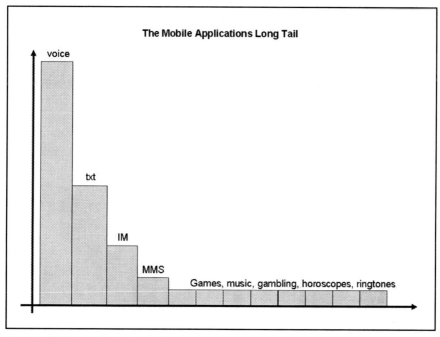

Figure 49: *Mobile applications long tail*

Here, the markets are fragmented and small, with segments of 10,000s or 100,000 users who are heterogeneous in their needs and consumption. The segments are so small because we, as individuals, differ in our consumption of the 'same' content. For instance consider stock market feeds: A stock broker may want real time access to market news and stock movement and may be willing to pay a monthly subscription fee. An investor may want the same newsfeed on a push alert paid for on a per transaction basis for the stocks she has invested in. However, an analyst may want the same information edited and delivered in a syndicated daily summary.

The service provider 'content' is the same but the proposition, target segment, method of consumption and value of each differs by consumer type. Operators are happy to provision and distribute 'off the shelf' content on their network, where the marginal cost of provisioning is minimal and cost of distribution marginal. On the other hand, Operators are reluctant to actively market or play the role of sales channel for a whole range of untried and untested Long tail content.

This fact is missed by many developers and content providers. Content providers and application developers often confuse these two aspects: distribution (fulfilment) and marketing (sales) when it comes to engaging Operators in commercial partnerships for their Long tail applications.

To an Operator, the collective revenue availed from all their data applications could be less than 1% of their core communications revenue, almost a rounding error! To compensate for this, content providers/developers have to sign up with multiple Mobile Operators to reach their audience. Thus, they to need be Operator agnostic. In conclusion, having a deal with an Operator(s) may not guarantee commercial success for consumption services and applications.

Demographics: Generation C (Gen-C)

The impact of Generation-C

The concept of Generation-C (Gen-C), where 'C' stands for community is discussed in detail in the book 'Communities dominate brands' [134] by Tomi Ahonen and Alan Moore.

---

134 www.communities.futuretext.com

What defines Gen-C?

According to Tomi and Alan:

- Continuous connection to digital communities and instant response to digital stimulus
- Very familiar with mobile phones and mobile communities
- Young (teenagers and young adults)
- Familiar with text messaging

Gen-C will shape the future because they will become mainstream consumers in the next five years. They are very familiar with technology and will drive the rapid proliferation of new technology. More importantly, from the Mobile Web 2.0 point of view, Gen-C is familiar with user generated content, as witnessed by the rise of MySpace. They are also more used to browsing and consuming content on the small screen. Hence, Gen-C would be a significant driver to Mobile Web 2.0.

Devices: Devices from a Mobile Web 2.0 perspective

A discussion about Mobile Web 2.0 and mobile devices

A mobile device is a complex topic in itself and whole books can be written about mobile devices. However, keeping our perspective of Mobile Web 2.0 in mind, we are concerned about devices from the perspective of:

- Capturing information at the point of inspiration
- Connectivity to the Web
- One device or many devices?
- The synergy between the device manufacturers and the operators

Capturing information at the point of inspiration : Increasingly, mobile devices are capturing non textual information: mainly pictures and video.

The Motorola RAZR V3X has a 2MP camera. The Nokia N93 records DVD quality movies. Onboard memory is increasing. The LG KG 800 has 128Mb of memory while the Nokia N91 has a 4G built-in memory. Expandable memory slots are becoming a common feature.

Connectivity to the Web: Not only can rich content be captured and stored on the phone, but there are more ways to deploy the content to the Web from the phone. The Nokia 6136 supports WiFi and 3G networks. A WiFi blackberry is also due in 2006. The Sony Ericsson K800i enables you to blog directly from your phone to the a blog site.

One device vs. many devices: The 'one device vs. many devices' issue is always a topic for debate. In other words, will people use only one general purpose device or will they carry multiple specialised devices (such as the mobile phone, the iPod etc)? Already the mobile phone continues to cannibalise devices: from the watch, to the MP3 player, camera etc. Our view is: the mobile phone will be the device most frequently used in any situation simply because it is present at the point of inspiration. This means, if you take 100 photos a year, a majority of them will be using a camera phone; but you will still take the camera for that special holiday.

The synergy between the device manufacturers and the operators: There is a power struggle between the network operators and the device manufacturers is ongoing. In most countries, operators subsidize phones. Also, they are the first port of call for customer services. This relationship will be tested by technology and emerging services. For instance, Mobile TV and video essentially provide a direct (broadcast) channel to the device which is independent of the Cellular network. Similarly, dual SIM devices like those developed by Eisar[135] in the UK are potentially disruptive because they shift the control to the customer.

To conclude, this discussion is by no means comprehensive but indicative of the trends we see in relation to Mobile Web 2.0 and devices.

Finally, an ironic point to note is: In a purist definition of Mobile Web 2.0, the

---

135 www.eisar.com

Ajit Jaokar and Tony Fish

browser becomes the lowest level of abstraction(and not the device). Thus, the capabilities of a browser could overcome some of the variations we see across devices if a browser becomes the 'least common denominator'.

Services: Ahonen's 5Ms

In our understanding of Mobile Web 2.0, let us not forget the first principles: What is a mobile service?

As the Web evolves and we create new services based on the principles of Mobile Web 2.0, it's important to remember the fundamentals of a mobile service. An especially important statement to remember is ***not all services can be 'Mobile'.*** So, we use Ahonen's 5Ms[136] as a template to identify a truly mobile service. The five M's are

- Movement: Escaping place
- Moment: Expanding time
- Me: Extending me (personalisation)
- Money: Expending financial resources and
- Machines: Empowering devices

These can best be explained by an example. On a rating of 0-5, the Mobile Ringtone service fares as follows with the 5Ms

---

136 Services for UMTS' (Wiley – Tomi Ahonen and Joe Barrett

Ahonen's 'M'	Rating	Comments
Movement	5	Because I can download a ringtone from wherever I want and it's simple
Moment	5	I can select the moment that is convenient to download the ringtone. If I see the ringtone promoted on TV, I can download the ringtone at that moment
Me	5	The ringtone is based on my tastes and dislikes and it says something about the kind of person I am
Money	5	I don't mind paying for the ringtone since it's quite inexpensive
Machine	3	When I download the ringtone, I do so interacting only with a machine (server)

Simple as they are, Ahonen's 5 M's provide a quick sanity check to identify a truly mobile service. We like them, because they force us to think from first principles. If a service does not lend itself to being 'mobile' no amount of marketing efforts will help!

Standardisation : The Mobile Web initiative from W3C

The impact of standardisation bodies on the design of Mobile browsing applications.

Ajit Jaokar and Tony Fish

Standardisation bodies

Because Mobile Web 2.0 lies at the cusp of the Web, Telecoms and Mobile networks, there are three sets of standards that we must understand when designing a Mobile Web 2.0 service. These are:

- **Internet standards:** Standards that govern the Internet world. The Internet engineering task force (IETF) [137]manages standards concerning issues such as packet routing, transport control and others relating to the core Internet itself.

- **Telecoms standards:** Standards that govern the air interface. Standardisation related to the air interface is managed by bodies such as the 'Third generation partnership program' (3GPP) [138]. The 3GPP is the place to go for all 3G, GSM and GPRS standards.

- **Open Mobile Alliance:** Standards that govern Mobile services. Open Mobile alliance [139](OMA) is the first port of call for all standards related to Mobile data services. OMA was formed by an amalgamation of a number of industry bodies such as the WAP forum, location interoperability forum etc.

In addition, other industry bodies that you may need to be familiar with include

- The GSM association : http://www.gsmworld.com/index.shtml

- The Mobile data association : http://www.mda-Mobiledata.org/

- Parlay : www.parlay.org

- UMTS forum : http://www.umts-forum.org

- WiFi Alliance : http://wi-fi.org/OpenSection/index.asp

- Bluetooth SIG http://www.bluetooth.org

- Symbian : http://www.symbian.com

---

137 www.ietf.org
138 www.3gpp.org
139 http://www.openMobilealliance.org/

- BREW : http://BREW.qualcomm.com/BREW/en/

- The Mobile marketing association : http://www.mmaglobal.com/

- Independent Committee for the Supervision of Standards of Telephone Information Services : http://www.icstis.org/

- Java is governed by the Java community process, which releases JSRs (specification requests). The URL for JCP is http://www.jcp.org/en/home/index

Also, we are now seriously entertaining the possibility of the 'One Web' i.e. a seamless Web spanning multiple devices and delivering the same information irrespective of the device that is used to access it. As might be expected, the concept of the 'One Web' is being driven from the Internet as opposed to the Mobile Internet. Specifically, in relation to Mobile Web 2.0, the standards bodies governing the idea of One Web are the same as the Internet standardisation bodies i.e. the W3C[140] (the World Wide Web consortium. The W3C is working with other bodies like the OMA to deliver a consistent set of recommendations for the One Web concept. We explain the impact of W3C documentation in detail below. Note that the W3C is making recommendations and not setting standards for the Mobile Web.

As per the W3C documentation: *One Web means making, as far as is reasonable, the same information and services available to users irrespective of the device they are using*. However, it does not mean that exactly the same information is available in exactly the same way across all devices. Some services and information are more suitable for and targeted at particular user contexts.

Within the W3C, the W3C Mobile Web Initiative [141]and further within it, the Mobile Web Best Practice (MWBP) Working Group[142] are working towards the goal of the One Web. At the time of writing this book, some of the standards and guidelines produced by this group are still under discussion. However, they will provide a good basis to design a mobile browsing application The

---

140 http://www.w3.org
141 http://www.w3.org/2005/MWI/Activity
142 http://www.w3.org/2005/01/BPWGCharter/Overview.html

Ajit Jaokar and Tony Fish

mission of the Mobile Web Best Practice (MWBP) Working Group is to develop a set of technical best practices and associated materials in support of the development of websites that provide an appropriate user experience on mobile devices. The Mobile Web Best Practice (MWBP) Working Group aims to extend the reach of the Web onto mobile devices by providing guidelines, checklists and best practice statements which are easy to comprehend and implement. When implemented by a website provider, these will enable the content to be perceived by users on mobile devices, particularly small-screen devices such as PDAs, smart phones and touch-screen devices. The MWBP Working Group expects to maintain contacts with groups such as the Open Mobile Alliance. Besides creating recommendations and best practises for the 'One Web', the W3C is also working towards the concept of the 'Mobile OK' trustmark. According to the W3C, the 'Mobile OK' trustmark will serve as the main conformance claim for the best practices document.

Impact on design

The above discussion impacts application design. Let us consider Mobile Ajax applications. By definition, mobile Ajax applications are Mobile Web applications. The impact of the preceding discussion is; when we consider the design of mobile Ajax applications, we have to consider it in two facets namely - design of a mobile website as recommended by the W3C and Ajax specific considerations as applicable to mobile devices.

Thus, to design mobile Ajax applications, we have to first understand the standardisation efforts by the W3C and then the specific factors relating to mobile Ajax. Also, note that the W3C recommendations discussed in this section pertain to site usability only. In a broader context, 'usability' could be defined as comprising of three parts namely: Site usability, device usability and browser usability.

According to the W3C definitions:

- Site usability relates to the structure, content and layout rules of a site and is a measure of the effectiveness of the mobile website.

- Device usability pertains to the capability of the equipment being used easily and effectively.

- Browser usability defines the ease of using a browser effectively namely performing the three functions reading, navigating or interacting. The ᴗᴀᴇᴗ ᴏf iɴᴛᴇʀᴀᴄᴛiᴏɴ, ᴩᴀɢᴇ ʀᴇɴᴅᴇʀiɴɢ, ᴄᴀᴄhiɴɢ ᴇᴛᴄ, ᴀʀᴇ iᴄᴄuᴇᴄ that are frequently used to judge browser usability.

In this section, we are discussing only the first part i.e. site usability since that is the only element defined by the W3C and under the control of the developer. In contrast, device usability is determined by the device manufacturer and browser usability is defined by the vendor creating the browser.

W3C design considerations
The W3C provides a comprehensive list of design factors and considerations. For a complete list, please refer to their website. In this section, we outline the general factors affecting the design of a browsing application. These include:

1) Presentation issues

2) Limited input capabilities

3) Bandwidth and costs

4) Keep user goals in mind

5) Be mindful of the impact of advertising

6) Be aware of device limitations

Presentation Issues:
Because web pages are created to be displayed on desktops, they cannot be directly presented on the mobile device in their original form. Not only is the overall user experience poor but also the content does not lay out as originally intended due to the different screen size.

Input:
Mobile devices have limited input capacity and it's hard to type in long URLs. In some cases, there is no pointing device (such as in some phones) and in general, it's hard to recover from errors.

Bandwidth and Cost:
Mobile networks can be slower than fixed line networks, they have a higher latency and in a majority of cases, the user pays for data retrieval. The device may support limited types of content. Thus, the user may download content only to realise that she cannot use it. In addition, the user may download content and may have to pay for additional data such as advertising. All these factors degrade the user experience and usability.

User Goals:
Unlike web users, mobile users have a very definite purpose/goal when they browse. In contrast, users on the web could browse for fun or to explore a topic (i.e. they are browsing without a specific goal). Thus, the mobile user seeks a specific piece of information and desire that it should be delivered in a format suitable for the device i.e. a short/exact response to the information request that can be rendered on the target device.

Advertising:
It is necessary to be extra vigilant when it comes to advertising on the mobile internet because the advertising could potentially hinder the user experience and it could not be free (because the user pays for the data download charges).

Device limitations:
Mobile devices impose limitations due to the screen size and limited input capabilities. In addition, there are other limitations due to the fact that there are restrictions on the software that can be executed on a device. In practise, this means – browsers may support limited or no plug-in or scripting capabilities

Further, some activities associated with rendering web pages are computationally intensive - for example re-flowing pages, laying out tables, processing unnecessarily long and complex style sheets and handling invalid mark up. Such compute intensive applications push the capabilities of the battery, memory and communications.

Conclusion

A number of standardisation bodies have to be considered in designing a Mobile Web 2.0 application; especially the recommendations of the W3C.

Applications: Mobile commerce, A barometer application

m-commerce is barometer for the acceptance of mobility In a society. People value what they pay for and the usage of the mobile device to conduct financial transactions is an important indicator of the maturity of the mobile market (which in turn is conducive to Mobile Web 2.0)

m-commerce

Mobile commerce (m-commerce) refers to the ability to conduct a commercial transaction over a mobile device143. m-commerce makes sense intuitively ; because a large number of purchases are made on impulse, and m-commerce is ideally suited for impulse purchases. Unfortunately, as with many other aspects of wireless, m-commerce was widely hyped. m-commerce is a great idea in principle, but difficult to execute in practise. In this section, we shall discuss m-commerce and the reasons why it has only been successful in certain markets.

Mobile commerce is driven by:

- **Ubiquity:** Anytime, anywhere purchases. For example, cinema, theatre or airline tickets

- **Time critical purchases:** Promotions and special offers

- **Purchase of entertainment type services:** Games and ringtones that can be consumed on the mobile device itself.

- The prospect of '**Micropayments**': i.e. getting customers to pay small amounts between 5p to £5. At these levels, credit cards are not viable because of their commission charges (normally a minimum charge of around 3%).

---

143 http://en.wikipedia.org/wiki/Mobile_Commerce

- The prospect of '**Mobile Coupons**': Magazine/Newspaper coupons have been used for a number of years but with the advent of m-commerce, coupons have become much more interesting. Coupons also fit in logically as part of mobile marketing. Typically, the coupon is sent as part of the marketing campaign on the mobile phone and is redeemed for a discount at the point of sale.

A Mobile commerce model ideally favours products that the user does not need to see, touch, feel or try on before purchasing, as they are in a standardised format. Examples might be books, CDs, DVDs, videos or tickets. Mobile commerce comprises two components: **Mobile Shopping** and **Mobile Payments**. Mobile shopping is concerned with the dialogue with the customer and helping the customer to place the order. Mobile Payments, on the other hand, deals with the issues of actually handling the payment such as security and credit card fraud.

m-commerce: The players

The market for Mobile commerce transactions includes a range of players each with their own strengths and weaknesses:

Mobile Operators are experts in managing networks and have a large subscriber base but lack the capabilities to develop compelling applications with speed and scalability. Mobile Operators also do not necessarily have the marketing and brand management skills needed to connect the right customers with the right products. Further, they do not have the capacity to handle payments (in the same way that the banks can). Mobile Operators are looking at Micropayments where they have leverage (typically, payments between 5p and £5).

The banks can handle payments but lack the technical infrastructure to handle Micropayments (unlike Mobile Operators). However, the banks are interested in this market since they see the Mobile Operators trying to encroach on their traditional business.

Credit card companies are interested in the market for the same reasons as the banks. Credit card companies are also interested in using their existing network more effectively.

Handset manufacturers have promoted the 'dual-chip' handset approach where the bank supplies a second SIM (Subscriber Identity Module) in addition to the mobile operator SIM that is then used for m-commerce. Currently, there is some focus on m-wallet technology.

Co-operative scenarios are collaborative efforts between the above parties for example: the Mobipay platform in Spain is available to all four Mobile networks. Other scenarios are possible - for example, between one mobile operator and one bank (e.g. Postbank and Telfort in the Netherlands) or one mobile operator and several banks (e.g. in Italy, the mobile operator TIM and the Italian banks).

Pure play vendors Such as Paybox who are solely in the business of providing a Mobile Payment platform. Back in the heady days of 2001, a study conducted by MORI, Britain's largest independent market research agency along with Nokia found that the number of people interested in using m-commerce is more than eight fold in some markets compared to the number of people actually using eCommerce today (2001).

From the study[144]:

> The study also showed that nearly 90% of people interested in using m-commerce services will also be willing to pay extra for the convenience of making purchases with m-commerce.

> The study revealed that Mobile-phone users view m-commerce as complementary to alternative remote-commerce channels such as the Internet.

> Specifically, the study found that initial adoption of m-commerce is likely to be on a similar scale as today's usage of eCommerce, which is already somewhat mature. Also, 24% to 54% of respondents across the markets stated they would be willing to carry out a transaction of more than USD 25 using a Mobile device.

> They tend to favour "local" transactions, where m-commerce provides a unique application for electronic transactions. Their choice of payment method depends on the size of the transaction and billing arrangements, but

---

144 http://press.nokia.com/PR/200105/822176_5.html

most are willing to pay extra for m-commerce services. This suggests that consumers see real value and benefit in using their Mobile phones as tools for shopping.

The study demonstrated that convenience and control will play a pivot role in acceptance of m-commerce. Study participants saw m-commerce as a way of avoiding carrying cash or waiting in queues, as well as a way to grain greater control over expenditures and enjoy unlimited purchasing possibilities.

In 2006, how many m-commerce transactions have you conducted? Not many...

What went wrong? Hype apart, m-commerce faces many hurdles: legal issues, privacy, willingness to pay extra for m-commerce, standardisation and so on. The present state of m-commerce is neither stable nor mature.

The biggest problem is the implementation.

In Europe, the last attempt to create a 'top down' m-commerce body ended in disaster (Simpay)[145]. But Simpay was only the latest disaster. There were many more previous attempts at m-commerce such as the Mondex trials with BT and dual chip phones as early as May 1999 in Helsinki. We shall discuss implementation issues in the subsequent sections and also see areas where m-commerce has been successful.

Implementation of m-commerce systems
Broadly, there are four ways to implement an m-commerce system

a) Through a **Mobile operator led initiative** like Vodafone m-pay

b) **Third party platform** led initiatives like LUUP

c) **Premium SMS**

d) **NFC (Near field communications)**

The overall issue hampering the uptake of m-commerce is the lack of a single standard. The direct impact of no ubiquitous standard is the biggest advantage

---

145 http://www.theregister.co.uk/2005/06/27/simpay_halts_project/

---

of m-commerce (i.e. impulse purchasing is gone!). The issues surrounding m-commerce manifest themselves in two ways: firstly retailers need to support the new method of payment and secondly, customers need to accept the method of payment and need to be able to 'transact with' ubiquitously (with anyone on any network).

In practise, in the case of Mobile operator led initiatives for retailers, it means you must be able to receive payments by 'Visa', 'Mastercard' and by Mobile operator initiatives such as Vodafone m-pay etc.

The same issue hampers third party initiatives. Consider the case of a company called Paybox (which still survives in a different incarnation today). Paybox approached m-commerce from the banking perspective reflecting their existing banking procedures (Deutsche Bank backed the company). Paybox connected the mobile phone to the bank. By making this connection it was possible to pay for services and products (and to transfer money to other people) using the phone. This is an example of the 'person to person model'.

To pay for a service using Paybox, users had to first register with Paybox.

Retailers had an integrated payment system with Paybox. When a consumer wished to pay a bill, the retailer entered the consumer's mobile phone number instead of their credit or debit card number. Then Paybox called the user's mobile phone to request a four-digit PIN to authorise the transaction.

Once authorised, Paybox informed the retailer, then debited the user's current bank account as with a normal direct debit payment. The user automatically received a text message as a receipt of the purchase, and a further email confirmation as an option. The retailer was also sent a message as confirmation of the payment authorisation.

With the benefit of hindsight, there were a number of problems but the most significant was that, retailers had to integrate into the Paybox system. It needed a significant body of retailers to make the system useful. Furthermore, how many such systems should retailers integrate into? At what cost; and to what benefit?

Ajit Jaokar and Tony Fish

Predictably, other initiatives came along. For example, in March 2004, Orange, the Mobile operator, ran a mobile coupon trial offering a 'two-for-one ticket deal' at over 90 percent of the movie theatres across the UK every Wednesday night. The problem in both these cases is the same. The retailer has to now integrate into a number of new systems each claiming to be the next 'standard'. The battle as we see below is at the 'vendor/retailer' first before we even get to the consumer.

Premium SMS

Contrast this with the only success on the m-commerce front: Premium SMS. The success of premium SMS was due to its simplicity and its ubiquity. Premium SMS involves a special type of SMS message that is chargeable. There are two ways to charge for a message, either on the outgoing message or the incoming message (Note – here 'outgoing' and 'incoming' are from the perspective of the Mobile operator)

Premium SMS MT (Mobile terminated)

The charging is done on the outgoing message. Premium SMS MT is preferably used when a return-SM is sent to the Consumer. Typical usage scenarios include applications where content is consumed or delivered within the SMS. Also, Premium SMS MT is highly suitable for subscriptions (one registration, many premium deliveries).

Example: Ringtone-purchase, subscriptions and alerting services.

Premium SMS MO (Mobile originated)

The charging is done on the incoming message.

Premium SMS MO is preferably used when a receipt or return-SM is not sent to the Consumer. Typical usage scenarios are applications where the content is not consumed in SMS.

Example: WEB-content, Voting/games without Mobile return-receipt.

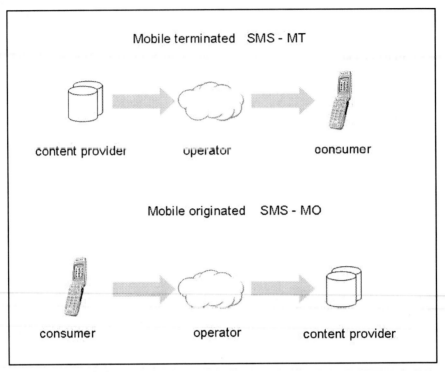

Figure 50: *Premium SMS*

The success of Premium SMS in many markets is encouraging but it is used mainly to purchase items such as Ringtones. m-commerce as a whole remains yet to be fully exploited.

P2P m-commerce

From our previous discussions, we have seen that excluding premium SMS, m-commerce has not had much success. The conflicting interest of the various parties involved is part of the reason for the slow uptake of m-commerce.

These limitations could be overcome if the market for m-commerce became homogenised i.e. anyone should be able to pay anyone else. Hence, we see a potential market for P2P m-commerce, providing there is a value proposition for the customer. For instance, the rise of PayPal[146] coincided with the rise of eBay because PayPal provided a mechanism to pay customers involved in online auction transactions. A similar 'driver application' for mobile P2P systems is yet to emerge. However, mobile P2P systems are already emerging. Two examples of Mobile P2P systems are: LUUP[147] and the mobile version of PayPal.

LUUP is a mobile wallet for online and Mobile Payments. The LUUP mobile wallet allows consumers to send payments to a mobile phone via SMS or online. Subscribers can also receive payments to their mobile phone from any bank or other LUUP user. You can send money to anyone who has a mobile phone from a country where LUUP operates. If the recipient is not a LUUP user, he/she will receive an SMS with the amount you've sent and an invitation to sign up for a LUUP wallet.

For example:

Tim owes Jennifer £6 for a cinema ticket

He knows her username is "Jennifer". He types a simple text message PAY JENNIFER 6 and sends it to LUUP's shortcode 81100. (Tim could also use Jennifer's mobile number instead of her username). Jennifer receives a notification of the payment in a text message. You can only send money from "cash" in your wallet (not from cards). You can transfer funds between a LUUP account and a registered bank account. Because the system works on shortcodes, it is available to anyone and does not suffer from the same issues as implementations from a specific provider.

---

146 www.paypal.com
147 www.luup.com

The other recent high profile announcement in the space of P2P mobile commerce is PayPal's announcement to support mobile transactions[148].

The P2P service works in the same way as LUUP above. However, PayPal also supports another service called 'Text to buy'. It works simply by sending an item code via SMS to a short code. A confirmation is sent by SMS and the item is then shipped. Due to its presence in the marketplace and its established user base, PayPal is expected to make a significant impact and also promote the usage of P2P transactions.

NFC – A New Hope?

Another more recent and promising technology that may overcome m-commerce's problems is NFC (Near field communication). Near Field Communication (NFC) is a short-range wireless connectivity technology that enables interactions among electronic devices, allowing consumers to perform 'contactless transactions'. In daily life, people living in London (Oyster card), Hong Kong (Octopus) and Japan (Suica) use NFC technology for cards in transportation.

The amalgamation of NFC technology in phones offers interesting possibilities. Devices containing NFC technology are triggered by proximity (a few centimetres). Think swiping your Oyster card on the London Underground. It's possible to equip a phone with NFC technology for as little as £3.00. NFC operates in the 13.56 MHz range.

Currently, there are two technologies covering this sector FeliCa[149] from Sony and Mifare [150]from Philips. The value proposition of NFC is to unite these two technologies. 'Unite' means to ensure that readers of FeliCa can read Mifare cards and vice versa. This could be VERY interesting. Obviously, the next step is to put in a mobile phone and all sorts of applications are possible. Interestingly enough NFC founding members includes players from across the spectrum, but the Mobile Operators are missing. Time will tell ... if NFC will be a success or yet another initiative which will not take off.

---

148 https://www.paypal.com/uk/cgi-bin/webscr?cmd=xpt/mobile/MobileOverview-outside
149 http://www.sony.net/Products/felica/
150 http://www.semiconductors.philips.com/products/identification/mifare/index.html

---

Ajit Jaokar and Tony Fish

Interestingly, NFC is more than a pure technology. In that, it has learnt from Bluetooth. Bluetooth concentrated on creating only the radio technology and left the applications to others.

However, it's not enough to create a technology like Bluetooth and have some start-ups create applications. In most cases, these applications remain in the prototype stage. It's far more important to have a major non operator consumer body behind the execution (such as the London underground).

In addition, NFC, in combination to RFID,(radio frequency identification) offers interesting possibilities for retailers especially when incorporated into mobile phones. Thus, apart from it's use in the transport network(such as Oystercard in the London underground), the technology is already being deployed in Hong Kong(via Octopus cards[151]) for a broad range of purchases such in apparel stores, bakeries, car parks, cinemas, food chains etc. NFC/Octopus is also being used in applications such as access control for apartment complexes and in school campuses.

Thus, companies developing systems based on NFC in Europe and North

America, are getting attention and funding(for instance Mobile Lime[152]), which secured $10m in March 2006.

In conclusion, NFC is an area to watch!

m-commerce: successes?

With so many twists and turns, one wonders if there is any place where m-commerce has been successful. Ironically, the most promising examples of m-commerce successes come from Africa. The lack of existing infrastructure (and consequently vested interests), is a bonus. A more efficient infrastructure based on m-commerce could be created in such a situation.

A company called Celpay[153] offers mobile phone-based virtual bank accounts

---

151 http://www.octopuscards.com/consumer/payment/use/en/index.jsp
152 www.mobilelime.com
153 http://www.celpay.com/zm/

---

with features like account transfers, bill payments, cash deposits etc. It has created a large retail base (shops, petrol stations etc) who will accept these transactions. (Celpay uses a solution from a South African company called Fundamo[154]).

Also according to the Feature[155]:

> Celpay has also developed successful m-banking business services. This includes mobile phone-based order entry with cash on delivery payment functionality. Current users include Coca-Cola, breweries and a cement manufacturer. In DRC (Democratic republic of Congo) alone, there were over 80,000 transactions per day on the Celpay system in November 2004.

Read that last statistic again ... 80,000 transactions per day! - in the democratic republic of Congo!!

There is another big boom in voucherless Top Ups. The agent simply enters the subscriber's mobile number, amount of credit needed and the agent's pin and the subscriber's account is topped up.

Of course, both Korea and Japan also show high m-commerce usage. In these cases, co-operation has worked in favour of m-commerce uptake as a whole. In Japan, Sony's FeliCa consortium has been far more successful in creating consortia. In Korea, the main operator (SK Telecom) is also an investor in petrol stations and this has helped in the deployment of their technology to the people.

In Europe, we see some companies like Scanbuy [156] using camera phones to take a picture of the barcode which they then use to send competing offers. Another company that has made some inroads in this space is Mobiqa[157]. Mobiqa focuses on concerts and events where there is a need for speedy interaction with large numbers of customers.

---

154 http://www.fundamo.com/
155 www.thefeature.com
156 www.scanbuy.com
157 www.mobiqa.com

Concluding remarks

We have spent a lot of time discussing m-commerce because we believe that it is a barometer application. In other words, societies and countries where m-commerce has taken off, truly value their mobile devices. In conclusion, m-commerce is still looking for its enabling application; it's 'eBay' so to speak (using the analogy of PayPal and ecommerce). We conclude with Buckley's laws of m-commerce, which we find very useful. These have been reproduced with the kind permission of Russell Buckley[158]

1. Usability

The overwhelmingly most important feature of a mobile payment system is usability for both the user and the merchant or retailer - I'll use "merchant" here to show off my knowledge of the sector. It's quite simple really. Here's Buckley's First Law of Mobile Payments:

If the transaction process is any more complicated than using a credit card or cash, it will never succeed.

2. Usability #2

It's not just a question of making the transaction painless from the user's point of view. It must be as easy to use as processing a credit card, by the merchant's staff.

Most retail organisations have a pretty powerful person called the Operations Director. This person's job is to ensure (among other things) that the shops work properly. As an aside, they've normally got where they are today by being good at shouting at people. Like all generalisations, this one is wrong too.

One of the key tasks of stores is to process transactions as quickly and

---

158 http://mobhappy.com/blog1/2005/06/15/mobile-payments-the-next-billion-dollar-market

efficiently as possible - and depending on the store, making this a pleasant experience. Think a busy grocery retailer on a Saturday before Christmas. If you add 30 seconds to a transaction time because your customer or staff are fiddling around with mobile phones trying to pay the shopping bill, you have a critical level operational crisis if this scenario is repeated.

Also, never, never underestimate the amount of time and money you will need to invest in retailer staff training. You'll have to tell them how to do this until you gag and still carry on telling them, while maintaining a serene smile.

3. Training

I know I already mentioned staff training, but the point needs labouring, frankly. If your expensively acquired new user presents her mobile phone for payment and gets anything less than "That'll do nicely, Madame" you've just lost that customer.

4. Simultaneousness

One of the problems we had when I worked on the launch of Air Miles (yes I am that old) was the chicken and egg. You can't acquire merchants until you can demonstrate that you have a critical mass of users. But you can't acquire users until you can show them that there's enough places to use their payment (or in Air Miles' case, enough merchants to collect from).

We managed it with Air Miles, but with a new payment system? It's a huge challenge needing significant and serious funding and some brilliant sales and marketing. The sales (acquiring merchants) and marketing (acquiring customers) must be symbiotic, which is frequently not the case. It's one of the few instances I can think of where having a Sales and Marketing Director might make sense, as this symbiosis is mission critical.

Recruiting users is much easier if you have a customer base already, like the operators and credit card companies already do. But, there's room for a well-funded start-up who can move fast and leverage some great marketing and cool deal making skills.

In this case, the chicken must come first. You have to acquire retailers first by convincing them that your multi-million dollar marketing campaign will acquire a serious slug of users. And that you'll help/pay them to convert their customers. And that you'll help/pay them with staff training and education.

5. Point of sale

A small point this and related to merchant acquisition.

When I lived in the UK, I once had a Diner's Card, which I never used to pay. Why? I simply never saw a merchant who had a "Use Your Diner's Card here" sticker.

Simple. But important.

6. Low cost

Credit card companies typically charge 5 -10% of the transaction value. Mobile Operators around 50% of Premium SMS revenue. You can't apply this level of rake-off and still remain competitive in most retail sales.

Therefore, the perfect system must have low/competitive merchant costs, or you'll never get anyone on board to take the payment.

Irakli, a payment specialist here in Munich adds this excellent addendum on his blog.

Security (Trust)

Security issues are still people's main concern regarding on- AND offline payment.

Security (Peace of Mind) #2

People like receipts. Although these are not technically necessary, people are used to be handed a receipt after a purchase in a physical store (different rules apply online).

Applications: The disruptive potential of Mobile TV

Mobile TV is a highly disruptive medium. It is much more than watching TV on your mobile device.

Introduction

Most industry analysts are bullish about Mobile TV. Driven by digitisation, broadcasting itself is undergoing a seminal shift, and these changes are flowing on to Mobile TV. In the UK, seventy percent of households have access to digital TV. That will jump to 100 percent by 2012, when the last analogue transmitter in the UK will be switched off. Digitisation brings choice. But it also changes viewing patterns because of features like the Personal Video Recorder (PVR) which can store hours of programmes. These programmes can be played whenever we want (i.e. content is now time shifted). In addition, the TV screen is not the only place where broadcast content could be viewed. It's also possible to view the same TV channels over the Internet (broadband).

Naturally, this logic can be extrapolated to mobile devices, and it is in this context that content owners are trying to deploy their content onto mobile devices. These devices could range from the iPod, the mobile phone, the PlayStation Portable etc. Broadly, there are two ways to transmit TV to mobile devices. The first is through the telecoms/Cellular network (**streaming**) and the second is through a broadcast network (**broadcasting**).

While there is always the potential to deploy video clips via streaming, as per an industry estimate, even if half of mobile phone owners watch even a few minutes of TV a day, 3G networks could overload within a few years. Source: analysys[159]. Thus, for most part, when we are speaking of Mobile TV, we are referring to *Broadcast Mobile TV* (as opposed to streaming Mobile TV).

Conceptually, 'broadcast Mobile TV' is similar to 'broadcast mobile radio' i.e. some older mobile devices have had the capacity to play FM radio. This is similar to any FM portable radio (and portable radios could be independent of

---

159 http://www.analysys.com/

Ajit Jaokar and Tony Fish

phones). The significant element being: the radio is receiving the content from the broadcast channels and not the Cellular network. Mobile TV also functions in the same way as mobile radio. And this has interesting possibilities, as we shall see below. Oddly enough, when the 3G forecasts were made back in year 2000, the potential of broadcast Mobile TV appears to have been overlooked.

Historically, the industry has always said:

- Text on mobile devices = GSM data services

- Still pictures on mobile devices + simple animations = 2.5G (GPRS)

- Video on mobile devices = 3G

When you factor in Mobile broadcasting, that picture now seems severely flawed, and with it, many of the market forecasts then produced. For one thing, the broadcast space is regulated by the broadcast regulators (which are different from the telecoms regulators). Thus, once again, the familiar spectre of spectrum issues raises its ugly head, even as the Mobile Network Operators have yet to recoup their investments from the 3G licenses just a few years ago. Not only could Mobile broadcast TV and video cannibalise existing 3G streaming video services, but they could end up introducing a whole new class of competitors from outside the telecoms industry (specifically players currently in the broadcasting industry or entirely new broadcasters). Thus, financially, Mobile TV poses serious question marks on a number of market projections and forecasts.

Mobile TV is FAR more disruptive than anyone realises

For example, once we all have the capacity to view video clips on handheld devices: which existing services will no longer be relevant?

For example:

a) Why would you want to view a picture (MMS) on a phone, when you can view a still image on mobile video (not to speak of the whole clip)

b) Considering there will be a limited amount of time when we are 'on the move' - will we play multiplayer games or watch video clips (thus cannibalising mobile multiplayer games)

c) And what about music? We are not sure of the stats on how much music (excluding ringtones) is downloaded on a phone via the telecoms network, but ... Why bother to download music, when you can get a music video?

d) Information services: everything from weather forecasts to stock tickers. Why go to a mobile portal, when you can get the same on a Mobile TV (broadcast)?

And so on ... In a nutshell, Mobile TV is definitely a major disruptive force. Let's now address the issues it raises in detail.

Standards

There are two notable standards for mobile video: Digital Video Broadcasting-Handheld (DVB-H) and Digital MultiMedia Broadcasting (DMB). There are others like MediaFLO160 from Qualcomm, but DMB and DVB-H are the two main players. While DMB is here today (most prominently in South Korea but also in Europe and North America), DVB-H requires additional infrastructure and spectrum availability. However, in Europe, DVB-H has greater mindshare and a number of ongoing trials.

DMB

The Korea Broadcasting Commission (KBC) and Ministry of Information Communication (MIC) coined the term Digital Multimedia Broadcasting (DMB). DMB defines technology adapted from the digital audio broadcast (DAB) system and developed by the country's Electronics and Telecommunications Research Institute (ETRI) during 2003. There are, however, two forms of DMB. Terrestrial DMB (T-DMB) which utilises a terrestrial broadcast network, and Satellite DMB (S-DMB) which utilises a combination of geostationary satellites and terrestrial repeaters. The latter method has been employed by TU Media,

---

160 http://www.mediaflo.com/

a joint venture between Korea's leading mobile operator SK Telecom (which holds a 30% stake in the venture), Toshiba and a number of other companies.

Both T-DMB and S-DMB utilises the video compression technology H.264 for video coding. In July 2005, the DMB standards received formal approval from ETSI.

DVB-H

DVB-H is an extension of the DVB-T (Digital Video Broadcasting-Terrestrial) standard for digital television distribution. DVB-H is spectrally compatible with DVB-T, allowing shared use of the DVB frequency bands with no impact on the performance of Cellular bands. In Europe, these extend from 470-862 MHz. DVB-H allows service providers to use the same digital broadcast infrastructure for both fixed and Mobile TV services. The network design allows service providers to address a variety of usage situations (outdoor or indoor; pedestrian or moving vehicle; etc.). Transmission of encrypted data is also possible. The DVB project [161]governs DVB-H but DVB-H is also strongly evangelised by Nokia.

The service

From a user perspective, the Mobile TV service involves more than just receiving the transmission on a mobile device. It could also potentially include:

- Live or on-demand reception
- Time-shifted Presentation
- Electronic Programming Guide, personalised by the subscriber
- Premium Channel Purchase
- Personalised Real-time and Stored Interaction, Programmed Alerts
- In-Stream Advertising and Mobile Commerce
- Integrated Music Content
- Integrated Location-Based Services,
- Unified Billing, Provisioning and Authentication

---

161 http://www.dvb.org/about_dvb/index.xml

Source: Mobile TV winners – Walter Adamson www.digitalinvestor.com.au, www.imodestrategy.com

From one minute 'mobisodes' for 'snacking content', to longer movie clips, there is a range of Mobile video content that can be viewed on a mobile device. However, inspite of the hype, it is actually unclear if people want to watch video/TV on their mobile devices. Customers have always listened to radio at work or when travelling. Would they be willing to watch video? Also, the usual culprits i.e. girls (adult), gambling and gaming do not lend themselves to this medium.

In a survey of 1,500 people, aged 13-55 for Olswang[162], only 17% wanted to watch television content on the mobile phone but 44% said they would watch programs on their PC. 70% did not want to watch TV on their mobile phones at all! Similar results were reported by Strategy Analytics[163], where fewer than 20% of the people polled (in UK, Germany, Italy and France) expressed an interest in watching mobile video. In our view, these reports are not indicative because Mobile TV is a much more radical idea, as we have seen in our service description previously, i.e. merely polling people is a waste of time with most radical innovations (which Mobile TV clearly is; as evident from the service description above).

The biggest positive uptake of Mobile TV is in South Korea. In South Korea Tu Media[164] , has already amassed 550,000 subscribers to its Satellite DMB based broadcast TV service, with phones running on SK telecoms mobile network (source Financial times London – print edition – June 1 2006). For Korea's nearly 50 million people and almost 40 million mobile phone owners, 550,000 Tu Media subscribers are 1.4% of the total phone subscriber base. In Korea, customers pay about 7 dollars per month to view TV content on their phones. On top of 550,000 on Satellite DMB, there are another 600,000 terminal (handsets, PDA, USB, laptop) sales for Terrestrial DMB, totalling to well over 1.1 million Broadcast devices for Broadcast Mobile TV.

---

162 http://www.olswang.com
163 http://www.strategyanalytics.net/
164 http://www.tu4u.com/index.jsp

Ajit Jaokar and Tony Fish

A comparison and player motivations

Here is a comparison of the two technologies and an idea of the motivation of the various players involved.

IP broadcasting:

DMB and DVB-H are conceptually similar since they are broadcasting IP packets to mobile devices.

Spectrum:

At the moment, DVB-H is still waiting for spectrum to be released by governments for example in the UK and other countries in Europe. Hence, we see many 'trials' but no 'live' installations. In fact, DVB-H is expected to be 'live' only around 2012 in the UK when the last analogue channels will be switched off. DMB has no such restrictions. This varies in other countries.

History of DAB:

There is already a large pool of DAB (audio) transmitters who could naturally take up DMB (video)

Channel support – national (DVB-H) and local (DMB) broadcasting:

There are technical differences between DMB and DVB-H but ultimately, the uptake does not depend much on specific technical differences. In fact, the two technologies can (and will) co-exist! In general, DVB-H is capable of supporting a larger number of channels, whereas DMB is probably more suited to smaller number of channels and localised broadcasting. Thus, we are likely to see Mobile Network Operators implementing DVB-H at a national level but with a large number of other players (broadcasters and emerging players) supporting DMB for localised broadcasting. (Note that the DMB trial for the world cup in Germany is conducted by T-Systems [165] and not T-Mobile i.e. a systems integration arm of an operator but not the operator itself

---

165 http://www.t-systems.com

DMB more disruptive:

DMB is the more disruptive technology because it enables a larger number of new and smaller players to enter broadcasting. Technically: retailers, department stores, existing DAB stations and many others could enter broadcasting via DMB because it enables localised broadcasting and the investment is not as steep as DVB-H.

Content providers and broadcaster benefit from controlling the value chain in DMB:

In a nutshell, they can bypass operators!

Tesco TV anyone:

DMB will allow more players to become 'broadcasters' - Tesco / Walmart TV anyone?

Handsets:

The key differentiator is handsets. In most countries, handsets are subsidised by the Mobile operator. Thus, to truly open up the market for DMB, we need cheap handsets supporting DMB (note they need not be phones!). In fact, we can almost bet that they won't be phones in case of DMB, considering that much of the operator community is supporting DVB-H. Thus, we expect to see many cheap DMB players (mainly Korean) to be the market enablers in Western Europe and North America. In contrast, DVB-H does not face this issue because most handsets will support DVB-H and will be subsidised by the Mobile Network Operators.

Operators:

Have already lost their leverage in broadcasting. They favour DVB-H mainly for its capacity to create a greater number of channels and it's large upfront investment (which inevitably keeps the broadcasters out of running a 'network' and implies that the operators are still the primary channel to communicate with the customer via DVB-H). It also provides Operators with a level of control

Nokia:

Is in a unique position of evangelising DVB-H globally and is the main beneficiary of the uptake of DVB-H in Europe and North America. Interestingly enough, for the first time, it is the dominant partner in the relationship with the Mobile Network Operators. Since its days of 'Snake' and the innovation of 'ringtones', Nokia has been an innovative company and has always had ambitions to play a greater role in the content value chain. These ambitions are well and truly realised with DVB-H. Overall, we see this as a positive development for the industry because Nokia has clearly demonstrated its innovative credentials.

Who supports DMB?

The Koreans? Good answer, but it's not that clear cut. The South Korean *government* actively supports DMB, both inside and outside South Korea. But that's not the case of ALL South Korean vendors. Samsung, for example, will merrily develop a DVB-H handset[166]

Governments:

Governments love spectrum! However, they are unlikely to make the same kind of gains as they did with 3G since the industry is wiser now!

The video iPod:

Is the potential dark horse? If the iPod model works (resynchronization to the PC), then video content may well be deployed through resynchronization with the web

User generated content: May well be the big winner with any standard.

DAB-IP:

Just to make life interesting, the world DAB forum which created DAB, also

---

166 http://www.samsung.com/PressCenter/PressRelease/PressRelease.asp?seq=20060214_0000233790

---

supports a parallel evolution path to mobile video **in addition** to DMB! They call it DAB-IP. So, from DAB, there are two parallel paths to go to video i.e. DMB and DAB-IP! As per the official press release of the DAB forum[167]

In addition to DMB trials, BT Movio[168] is also running trials for DAB-IP. This is confusing for most analysts and initially we totally missed out DAB-IP as well and confused it with DMB, but the two are definitely different!

MBMS:

Some telecoms operators are using MBMS (Multimedia Broadcast Multicast Service) as an interim solution to broadcast over existing GSM and UMTS Cellular networks.

Conclusions

a) As you can see, the market reality is complex and it's more than a pure technology issue: DMB and DVB-H. The reality is more complex as we have seen, especially if you include the video iPod and DAB-IP. But, whatever the combination, we believe that these technologies will co-exist: **this is not a GSM/CDMA situation where there is only one winner**

b) We believe that DMB is more disruptive than DVB-H (mainly because DVB-H is suited for larger players since it needs an initial significant infrastructure investment). DMB will be suited to the many existing and new content providers (partly because dealing with operators is difficult and the content providers can control the whole value chain

c) Consequently, of course, DVB-H can support a larger number of channels. Thus, we see DVB-H to be deployed nationally and DMB for regional broadcasting

---

167 http://www.worlddab.org/images/DABMobileTVat3GSM.pdf
168 www.btwholesale.com

Ajit Jaokar and Tony Fish

d) Our reasoning for the conclusion (all technologies will co-exist, DMB will be more suited for regional broadcaster and DVB-H for national broadcasters) is based on the capacity to support channels.

DMB (and also DAB-IP) support a very small number of channels (in some cases as low as four). But ironically, that opens up new markets for DMB

Ignoring multiplexers, boosters and other signal enhancing techniques: considering the example of terrestrial DMB (and not satellite) for Germany. Germany has a total bandwidth of 1.5Mhz. It can optimally support four simultaneous channels. More channels could be possible if quality were less stringent and we used other means like multiplexers etc, but optimally four DMB channels are supported in Germany. Korea itself supports 7 channels nationally on terrestrial DMB. Thus DMB could go national, but the number of channels supported in Korea is 7.

DVB-H on the other hand, needs a much bigger infrastructure: hence needs more channels almost from the start to make it viable. This could be either good (more choice) or bad (more choice!!). So, typically, DVB-H has a bandwidth of 5Mhz to play with. Thus, you could fit in more channels if you have to but the point is, due to the infrastructure investment, you **need** those more channels to make it viable in the first place.

e) In either case, Mobile TV and video remain a disruptive technology

Applications: The disruptive power of Podcasting

Like Mobile TV in the previous section, Podcasting is also a highly disruptive medium and will have an impact on Mobile applications.

Introduction

Although digital audio has been possible for a while (e.g. VoIP, DAB, IP streaming), a spate of recent developments over since 2003 has shifted the landscape significantly. Principally, these are (i) The increasing penetration of broadband (bandwidth allowing reasonable quality audio to be guaranteed), (ii) the use of peer-to-peer and other web client based technologies, and (iii) the increasing convergence of fixed/mobile technologies. This has already driven some major changes to established industries:

- A rapid rise in VoIP adoption.

- The proliferation of music in digital form; initially available in dial up services, its popularity started with Napster and then peer to peer (P2P) networks. Music over IP has grown rapidly since about 2002.

- Radio over IP has started to grow fairly rapidly over the last 3 years.

Most recently, the phenomenal growth in the use of portable, personal digital media players (MP3 players, and more recently Apple's iPod) has added the mobile element to digital audio. Combined with the Broadband Internet and the "Blogging" mindset, this has led to a new type of broadcasting to mobile devices (iPods), hence "Podcasting" (which is not necessarily confined to iPods). According to some, this is just the beginning of the impact of the broadband audio disruption - Broadcast Radio, 3G Mobile Telephony and broadcast based advertising are all said to be in line for major change. Therefore understanding Podcasting is important, however, not just because it is a potentially disruptive technology in its own right (Radio is an £20bn+ industry globally after all), but also because:

- It is one of the most visible of the services being driven by the changes to the way the World Wide web and the Internet operate i.e. Web 2.0

- The structure of the audio industry is similar to that of the much larger video industry, so lessons learned here will be applicable to video as bandwidth increases.

Podcasting has a number of features in common with other disruptive IP media technologies, in that it:

- is capable of time-shifting, i.e. allows listening when the customer wants it

- allows ad-shifting (avoiding unwanted adverts)

- is global - access is via the global Internet, not the radio station or satellite

- is user-programmable, i.e. the customer selects the content they want – drives the move away from pre-selected, more mainstream content

It is also potentially more disruptive than other IP based audio, as the technology it uses is:

- a fixed/mobile convergent technology, allowing portability

- using new push type software that potentially allows each user to aggregate the content of their choice, disrupting the portal industry

Podcasting also has a number of impacts on the legal and regulatory framework in that:

- it is by and large an unregulated medium, so a "Wild West" environment is likely in the early days

- status of rights, royalties and other copyright mechanisms will no doubt be tested to the limits

Podcasting – A brief overview

In brief, podcasting works by creating or digitising content into an audio standard (e.g. MP3), and using the RSS (Really Simple Syndication) standard to transmit content over the Internet to a customer's PC or device. That PC uses "Podcatcher" software to find, store and play the audio "Podshow". It can also transfer the podcast to a mobile device. Customers search generic or specialised websites to download the content onto the PC, and onto the Pod.

Podcasting was initially named by the early blogging enthusiasts using new technologies to broadcast over IP to Apple's iPod. Podcasting is now the generic term that describes using PC or mobile personal players to download and store audio broadcasts, initially music and increasingly other types of audio such as radio, audio books etc, and has spawned a new grass roots "podcasting" industry like web publishing and blogging did before it.

Figure 51 shows the main technologies used in the Podcasting supply chain stream, and their approximate costs of acquisition and use.

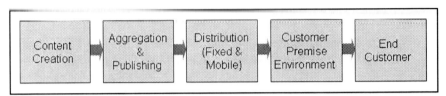

Figure 51: *Podcasting value chain*

Is Podcasting really a Disruptive Technology?

Just as Blogging has made journalists of previously ordinary people, will Podcasting make us all talk show hosts and media personalities? Will this damage the business models of the more established audio industries? Podcasting has one of the fastest rates of adoption of a new consumer technology ever, even the name shows the disruption...the term Podcasting coming from Apple's iPod, which burst on the scene from nowhere in 2003, sales rising 600% in one year. It is the integration of the Blogging mindset, the personal digital media player, and broadband Internet penetration.

Consider that:

- The number of "Blog Sites" has risen from a few thousand in 2002 to c 60m in 2005

- Apple shipped 6.2 million iPods in Q2 2005 for $1.1bn – a 600% increase from 2004

- In the US, c 22m people own an iPod/MP3, 29% of them have used Podcasts. That represents a c 5% penetration of total population.

- There were 526 Google hits for "podcasts" on September 30 2004, then 2,750 three days later. The number doubled every few days, passing 100,000 by October 18. After only nine months, a search for "podcasts" produced more than 10 million hits, and as of September 2005, the same search produces 61 million hits.

Ajit Jaokar and Tony Fish

Blogging, VoIP, and now Podcasting, are considered to be some of the most visible parts of the move to Web 2.0 which could be seen as the real time, fully interactive, grid enabled, broadband internet. Many of the new features of the Web are driven simply by bandwidth and penetration increase, and some of the underlying behaviour has stayed remarkably constant over the last 10 years, as Figure 52 illustrates.

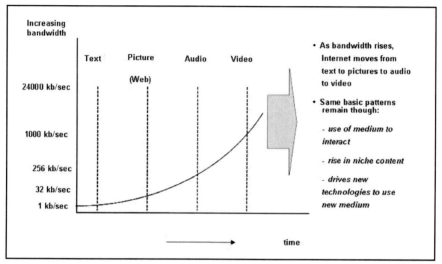

Figure 52: *Lessons from Internet Adoption as bandwidth increases*

Impact on existing audio businesses

From basic text in the early 90's through the evolution of the Web, Blogging, and now Audio, a pattern emerges across the supply chain of the Internet's propensity to:

- The medium is used to connect people with common interests, across national and global boundaries.

- This drives a rise in the consumption of less popular niche content – the "Long tail". As transaction costs fall, so smaller communities of interest can be served with even more unique content…drives a rise of niche yet global communities.

- Unlock creative talent among "enthusiastic amateurs" globally to serve this demand. A wealth of new content emerges.....and often illegal redistribution of existing material if copyright or access restrictions are onerous.

- Drive entirely new aggregation and publishing businesses, frequently produced by smaller, new players capable of rapid growth.... e.g. Yahoo, Google, Skype.

- This leads to a number of assaults on the existing businesses once a "tipping point" of usage is reached.

- The 'Net' attacks existing distribution networks with its (usually) more effective digital packet switched technology (anyone remember SNA, ISDN....).

- It attacks existing end-to-end channels with a lower net cost technology. The continual evolution of simple, open standards, (RSS - Really Simple Syndication in Podcasting's case), typically drives continuing cost reductions as well as driving increasingly merged multimedia capabilities.

- The combination of lower cost technology and ease of distribution drives lower transaction cost, which allows 'Net' based technologies to attack established business models.

- Advertising assumptions that usually underpin most traditional media are altered, thus impacting the business models.

The net effect is one of both directly challenging, and of unbundling, traditional players' business models (the in-word was disintermediation in the 90's) to the extent that new players can enter into parts or all of the supply chain to challenge them. The major issue in the early days for any incumbent is that the new models often have negative impacts on their business, before any money is made per se – making it hard to justify entering the new market early. Soon, however, the new service appears in strength, competing with traditional media for the time, wallet and mind space of consumers everywhere. Skype (and VoIP in general) is a very recent example.

Impact on Video

There is also a strong possibility that the Audio world will provide a prediction for the impact of broadband on the Video world. This is partly because (i) the Audio world is adhering to the patterns of the previous IP based communications mediums (text, web), but also that (ii) the structure of the audio industry bears a strong resemblance to that of the video market. In fact, early peer-to-peer DVD copying is occurring in imitation of the impact on music. Thus, the lessons honed by entrants in Audio may well be used later in the Video markets.

But will it really happen – we've been here before, after all?

To be truly disruptive however, a technology must radically change the status quo, create anarchy in the existing industries and supply chains, and turn old advantages into liabilities. It must be able to destroy large incumbents and create new giants. It must change the way people live and companies do business, and create shocks in the regulatory and legal frameworks. Can Podcasting really do all this?

Looking at previous technologies that have proven themselves to be genuinely disruptive, a number of general principles can be drawn:

- Good enoughs - the technology must offer a significant improvement over existing "good enoughs" to be attractive to all but "early adopters".

- Ease of Use - It must have features that allow it to be easily and rapidly adopted by the majority of end users. Ideally the end user already has the components, or can be persuaded to acquire them – failing that; they have to be given them by the operators.

- Significantly undermines existing business models used by the major current players, or it will merely be absorbed by them, rather than spawning new companies.

- Sustainable business model of its own that allows the technology to grow past the early adoption stage. If the technology does not have a business model, it will not gain external funding or momentum.

- It is not enough to attack just one part of the supply chain; it must threaten multiple areas to make assimilation difficult (and entrance by outside players attractive).

Taking these in order:

Sufficiently better than "Good Enoughs"?

Existing "good enoughs" are the current audio services offered. There is significant evidence from previous rapid technology adoption that communication facilitation, portability, ad-shifting, time-shifting and lower costs are constantly attractive to users.

Recent research suggest audiences do not like the amount of advertising on radio, want to shut up the endless rap of disc jockeys, and would like to have a larger say in the content. This is mainly US data admittedly, but the UK is not usually too different.

The time shifting issue has already raised its head in the Video industry via TiVo and other Personal Video Recorders, and it's hard to believe that a similar Audio effect will not occur, but it is unclear to what extent it will penetrate as audio tends to be consumed as a "wallpaper" to other activities, so ads are not as intrusive. There is also emerging evidence that audiences do not like the bland "common denominator" programming that radio stations – Business Week recently reported that listeners are increasingly tired of homogeneous radio programming, as there is little uniqueness. What is unclear is how much interactivity is valued, but most experiments have shown that interactivity is attractive, even if its just of the "have your say" variety.

Easily adopted by end users?

The more that the new technology can use existing infrastructure, typically the faster its rate of introduction. Podcasting uses a lot of existing infrastructure, and Indications so far are that it is growing rapidly. However, two barriers still stand in its way to mass adoption:

Is it easy to set up and use? Not yet... tech-savvy early adopters can do it, but our experience from more mass-market customers is that they still find it hard to set Podcasting up. It still needs a transfer from PC to Pod that means effort… and the mass market does not like effort.

Who manages the Metadata? One of the reasons iTunes has been so successful is that the database is easily searchable using a fairly simple tagging system, and uses a format common across the music industry (CDDB). In fact, herein lies a tale, CDDB was originally a shared resource that went commercial after acquisition. This has so enraged the developer community that a new freeware standard – FreeDB – is now being built by music enthusiasts globally.

It is not clear what audiences Podcasting will eventually reach - the whole radio listening public, or just niches? Will it be aural wallpaper for people on the move (i.e. more likely competing with other music products), or listened to more intently, allowing other voice based services? Figure 'Potential usage of Podcasting' gives one view of potential application areas – in theory the lower right hand square - low competition, high popularity is the best place to be.

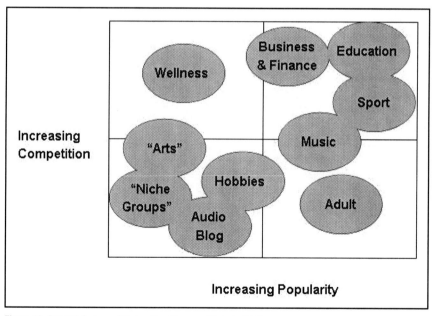

Figure 53: *Potential usage of Podcasting (Source – Verkata LLC)*

Undermines Existing Business Models?

Figure below is a simplified representation of the media supply chain, showing the major functional areas and the impact of podcasting on them. As the diagram shows, digital audio is impacting across the value chain.

It is interesting to look at potential impacts along the supply chain from content production to consumption to understand where the major disruptions are occurring, and what the knock-on (and knock-back) effects potentially are. The figure below describes the impact of Podcasting along the digital supply chain.

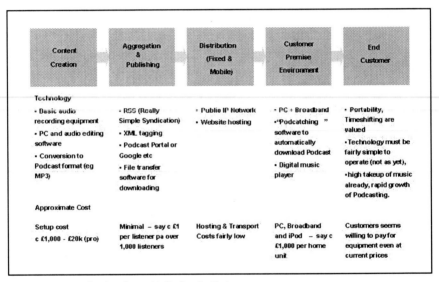

Figure 54: *Impact of Podcasting on Media Supply Chain*

Podcasting thus has a number of common areas that are potential causes of major disruption across the supply chain.

Content Creation – The sourcing of the conception, design and production of the content. The main dislocation is the rise of new creators, using low cost modern technology to bypass the current industry structure. The lesson of the Web and Blogging is that this will occur in the Audio space, mainly "Narrowcasting" to distinct interest groups.

Ajit Jaokar and Tony Fish

Content Aggregation and Publishing – The process of sourcing, selecting, marketing, and rights management of the content, also the billing of the content itself. The rise of new technologies gives an opportunity for new aggregators to reach their new audiences, at the expense of existing players. These can be legal newcomers, typically smaller start-ups or entrants from other industries, or illegal operations such as peer-to-peer networks. There is also a lack of existing tagging/metadata leaving the field open to new sequencers. In addition, Podcasting seems to be driving systems that aggregate at a user, rather than publisher level.

Distribution – The method of transforming and transporting the content to the end user, and any separate billing for the transport if applicable. The fundamental shift here is using broadband IP to bypass 3G mobile, DAB, Cable media, traditional telephony and hardcopy methods (e.g. CDs, Audiocassettes)

Customer Premise Environment – The iPod, PSP, PC, Mobile Phone and potentially PDA can all do some combination of communication roles, all can do Audio, and it may increase the utility and attractiveness of one method over another, tipping current behaviours in new directions. To be multimedia capable, they all seem to need the PC and broadband connection – which plays to the IP network's hand.

A Sustainable Business Model of its own?
Digital Audio has a number of potential impacts on this value chain. It is clearly already:

- Capturing New Users.

- Shifting the role of Advertising in the value chain.

In addition, it clearly reduces setup costs of an audio broadcast service.

- It is not paying for audio spectrum, 3G licences or cable buildout.

- The basic technology required is much cheaper than most broadcasting technologies used today.

- Low cost of user subsidy – the user buys the "Podman" and computer, it does not seem necessary to give them a set top box of any sort.

- Similarly, there is no "truck roll" to set up the systems.

From a CAPEX point of view, the main impact is that Podcasting uses standard Internet/ Hosting infrastructure, so is less likely to require ongoing specialised infrastructure spending, thus again impacting its overall competitiveness. However, to make money Podcasting still has to show it can increase revenues from today's low (near zero) base, and reduce churn when the next "new *new*" thing comes along. Options being experimented with today are:

- Advertising – Virgin has advertising in its Podcast service, and Yahoo has tried 30 second clips every so often on its Radio service. Most pundits believe that advertising has to be narrowcast to the user to succeed. There is even a view that advertising will have to "nanocast" to tiny niches to work. However, others feel that users will not have the inclination to download an advert and then upload it to their player.

- Sponsorship – along with advertising this was the staple of early commercial radio and is being tried, for e.g. Volvo on Autoblog. Again, sponsor probably needs to be correct for the niche.

- Subscription –many hope that podcast audiences will follow the subscription model, but it's early days. The history of the 'net so far is that unless people deeply desire the content, subscription is unlikely to take off.

- Micropayments – many websites, blog sites etc already allow voluntary donations via PayPal, it remains to be seen whether users will actually pay for content.

- Subsidy – the BBC already puts a lot of content onto podcasts, and this is subsidised via listener's fees. Other companies will be tempted to podcast for free, hoping to collect benefits via higher customer capture levels for other services, increased "stickiness", or just the marketing benefit of looking "cool".

Some even believe that it will have to appeal to extremely small audiences to work, i.e. to "nanocast": "Nanocasting aims to deliver programming profitably to very small but highly interested audiences. Using radio effectively in nano markets requires very different strategies and opens the door to programming that would have been unthinkable and unsustainable under the broadcast model" (International Nanocasting Alliance, 2005)

The future of Podcasting – Three Possible Scenarios

One way to look at the possible disruptive outcome over the next few years is to describe scenarios of what may occur, and then look at what one needs to believe for those scenarios to come about. For the purposes of this study we look at 3 possible scenarios – Absorption, Assimilation and Assassination. They are sketches rather than fully worked through scenarios, but they should give some food for thought.

- Absorption – Podcasting technology is absorbed by the current industry to create new services and add to existing ones.

- Assimilation – the technology allows a number of new companies to emerge with new services, and they join the industry sector serving the customer base. There is some reduction in the prospects of the existing players, but no threat to their existence

- Assassination – the technology not only allows new companies to emerge, but their business models destroy value for existing companies and they take large amounts of customers (and spend) away from them.

In all these scenarios we assume an early growth phase occurs, with early adopters taking up the new services warts and all. The issue is what occurs at the "tipping point" - the point at which the service can potentially jump into the mass market (typically about 10% of all users, i.e. the US situation fairly shortly).

Absorption Scenario

Podcasting software becomes simpler, and it jumps out of the "early adopter" market as large numbers of people start to use it as a new way of accessing

---

content - driven by friends' recommendation and viral marketing. This starts a blooming of small, creative, content producers and new stars are born – some to become new media favourites, many to burn out shortly after. In the early days there is much hand wringing as traditional companies' audiences dwindle, but the spend fails to move away as fast. Some existing companies take to the new medium – Waterstone's Podbooks service (bought as a small start-up) rivals Apple's iTunes, catching Amazon on the hop.

However, customers show that they are fairly unwilling to pay much money for many of these new services, and any attempt to make them pay leads to rapid migration to other emerging free services - the industry has few barriers to entry so a new service is always opening up, taking customers from existing ones. In this scenario, absorption occurs in a number of ways. Large portals such as Google and Yahoo use their portal strength to reproduce the emerging new services, even in some cases to buy them. Commercial broadcasters put a large amount of their content onto Podcasts for free initially, later becoming advert supported (subscription only works for adult, business and horoscopes).

Microsoft reproduces many of the basic Pod technologies in the next Windows release. In the UK, the BBC gets behind Podcasting in a large way, and because it is largely free it becomes hard for any small commercial undertakings to make any money, prompting angry letters to the Regulator, the CBI and the Times. Initially, advertisers are downbeat as users embrace the new technology to avoid any "message from their sponsors", but slowly start to use the more focused nature of the service to "narrowcast" their messages. In the early days, this leads to some interesting results as Podverts aimed at more urbane European audiences are taken up by more conservative people in Asia and the US. The music industry continues to struggle with piracy, but method pioneered by players such as iTunes prove to be the way forward for music Podcast as well, and piracy slowly reduces as a major threat.

By 2010, Digital Audio is a multi-media service offered by most of the existing players in the value chain, Podcasting is one part of it. Some of today's companies have done a bit better out of it, some a bit worse, but otherwise its business as usual.

Ajit Jaokar and Tony Fish

Assimilation Scenario

The start of Digital Audio looks much like it does for the Absorption scenario, but even in the early days a number of other underlying trends become apparent: Customers are willing to pay for small downloads and subscribe to specialist services, and revenue leaks away from existing services such as ringtones as Podcasting is the "new *new*" thing. Advertisers rapidly adapt to narrowcasting, and even "nanocasting"

This monetisation allows the many new content creators and services to stay independent and attract new talent, and increasingly Venture Capitalists are attracted to funding new businesses in the area, generating a boom in exciting new content. Young people drop spending on mobile services such as ringtones, and the mobile is relegated again to a mainly comms device, as the broadband content is consumed via the new Podman products coming out. There is a movement away from Web surfing as a recreation of choice towards consumption of audio material as people find the "audio wallpaper" allows them to get on with other tasks.

The IPO of a pod metadata business starts a mini rush of funding into the sector as VC herd behaviour re-emerges. This investment allows some small companies to rapidly grow to (un)sustainable size, but sidelining existing Portals in the audio (and later video) space and adding further disruption into the market. Every household already has a PC/WiFi environment so podcasting becomes a major way of taking audio content, especially for the younger members of the family. There is huge pressure on regulators worldwide, as claims of piracy, offensiveness and corruption of minors from the new media are rampant.

The BBC, under pressure both for wasting UK taxpayers' money to serve customers across the world, and for spoiling the UK market for innovative podcast companies, reduces its efforts – further increasing the space for new players.

By 2010 a new order has emerged, a number of new champions have barged their way into the market, some have burned out – taking existing competitors with them – and the telecoms industry has seen a fall in usage of its proposed IPTV and 3G services. Major portals have taken a blow – but not a mortal one – and eBay has bought PodCo as a strategic move.

Assassination Scenario

This scenario starts off much like the Assimilation scenario, and the established industry is reluctantly accepting the brash newcomers muscling in and grabbing some of the pie, when 3 other events occur. VoIP growth has driven a major rise in low cost WiFi pickup points, where users can access the 'net for their calls in a near-mobile way. Emerging metadata tagging means Podcasting is the de facto way of sending any audio "clip" up to about 30 minutes in length. There is a technical change to the "PodPhone", which can now receive text and audio…and send it back. Podtexting, Podmail and Podcalls become the new Unified Messaging Service, initially for the teens and twentysomethings, and a number of (well funded) new companies rush to market to exploit this trend.

A new service is released, where gamers can talk to each other while playing, and send Podcasts of guidance to other players. This drives the creation of an entire new set of services based around multimedia services via a much more user-friendly device than the PC, which captures a huge amount of the time – and spend – of many of the core spenders for existing services as well as many women who up till now have not really participated in the "on-line" world. The main content houses find that the cost of re-formatting their content to the new interactive media format is very high, delay doing so, and are thus put into head to head competition with new content creators. While their marketing muscle helps, the emergence of automated pull based EPG's means users can find lesser-known content they desire fairly easily on the Web without any intermediaries.

Simultaneously the rise in illegal P2P video file sharing impacts revenues, and the formation of an industry consortium, iVid, is too late to stop the decline. DAB is put on ice as it becomes clear that the UK will take its digital radio via PCs and cheap pod devices hooked up to the net. The net impact is major disruption across the industry supply chain. Existing businesses revenue streams decline while costs of customer retention rocket, and the huge amount of money sitting in the VC and private equity markets is released to fund the upstart industries. PodCo's MediaPortal becomes the clear winner in the multimedia navigation stakes, and its user recommendation model proves superior to any search-based model.

Existing Portals' valuations fall by an order of magnitude as it becomes clear that search is no longer an intermediary's role for audio and video content. The Mobile industry is badly impacted when it becomes clear that much of the broadband audio and video service expansion will pass it by.

A closing view

Knowing that prediction is risky (especially about the future) nonetheless it seems that based on the evidence so far the Assassination scenario is less likely than the other two. This is not to say it is not possible, just that events that probably need to occur have not done so - yet.

From past experience, the response of the current players will determine how much time and market space the Assimilation scenario has to play out - ignoring the emerging technology is more likely to bring it about. Have the current players learned their lessons from the past?

Some are embracing Podcasting, others are ignoring it. What is very likely to be true is that the Podcasting industry will cause a number of shocks to some big name existing players, create some new household names - both as broadcast stars and as new companies - and generate a lot of media attention in the process. This will make it attractive to the younger set who will migrate away from the current "cool stuff" (e.g. ringtones), leaving a number of sub-sectors facing decline.

This potentially suggests that one outcome is an eventual Absorption scenario, but at higher prices as the existing players buy up emerging yet valuable Podcast companies at Assimilation level maturity – and prices. What is very likely is that Podcasting will ask some searching questions about the legal and regulatory environments under which media currently operates, and force some uncomfortable changes to the status quo in these areas.

Applications: I am not a number; I am a tag : implementation

The implementation of 'I am not a number, I am a tag' (the second principle of Mobile Web 2.0)

Introduction

Previously, we have discussed the idea of 'I am not a number; I am a tag'. In this section, we shall discuss the implementation of this concept. To recap, we are discussing a new form of search engine: a search engine of (verified) tags. Here, tags have the same meaning as they do on Flickr etc except that they are verified and they could connote either businesses or people. This search engine is driven by an (as yet) hypothetical internal algorithm which we call for the lack of a better word, 'PeopleRank' (analogous to Google 'PageRank').

The basic idea is 'tags' could be used to make calls from the Web to any number.

A directory based on an algorithm (which we have called 'PeopleRank') would act as a web based user interface (and in effect a virtual operator) and place the calls. The concept is based on convenience to the end user. When Amazon was first launched, many people thought Amazon was about cheaper books. Today, we don't necessarily expect cheap books from Amazon; but we get 'something more' i.e. the choice of a range of books. Similarly, with 'I am not a number, I am a tag'; you get all the numbers in one place, you get presence information and you can call the numbers from the Web.

In a nutshell, PeopleRank indicates a complex algorithm for classifying people/businesses (a directory) but it also incorporates their profile/presence information. Thus, the PeopleRank directory/algorithm knows who you are, the context (for instance; where you are) and how you wish to be contacted. Such a mechanism is driven from the Web and works at the application level of the stack (in other words, it is not 'fixed to mobile convergence'). It takes away the locus of power from the Mobile Network Operators and back to the user through a profile.

Ajit Jaokar and Tony Fish

The idea of 'I am not a number; I am a tag' sounds enticing enough. But how would it be implemented? Let's break down the functions:

a) Profile: We need a mechanism which stores user information including context (location, presence etc) i.e. a profile.

b) Identity: We need a trusted identity i.e. the ability to reliably identify the person and their information when we want to call them.

c) Addressing: Finally, we need the ability to discover and address the mobile device independently of a telecoms operator (else it defeats the purpose and the idea no longer becomes disruptive). We call this 'bridging the wireless gap' i.e. the last wireless mile.

Profile
The profile requirement is the simplest. Already, many social business networks maintain a profile. That profile can be used to store a range of contact information which may be made public or released to a selected audience. Presence information could be manually updated or could be set via rules.

Identity

Introduction
Identity is a more difficult concept and needs a lot of thought. Today, governments can increasingly identify us from biometric information. This kind of identification arises from primary sources (biometric) and is validated by organisations which are accepted and recognised globally (the US and UK governments). Hence, my primary identity is defined by these primary facts (biometric information, birth certificate etc). This leads to conferred identity (secondary identity) such as a driving license or a passport. Providing the issuing authority of these documents such as a driving licence or a passport is recognised and accepted, the secondary identity is also as good as my primary (biometric) identity. Of course, there is a risk that the secondary identity may be forged (such as fake passports). That risk, coupled with increased global threats, has caused many governments to revert back to primary identity (retina scanning/finger prints). These of course, cannot be forged.

Identity in the digital world

On the Web, identity is still relatively primitive. Identity is what a system knows about me at the time of registration. This may not be true. Every time I register with a new system/site, I create a new identity. None of these identities are tied to any biometric information (currently). They are silos because each site has its own membership system. The obvious way to solve this was the 'single sign-on' method. Microsoft attempted this through their passport mechanism169. But that did not receive much response. Initiatives like the liberty alliance 170are a consortia of companies. But consortia often have an important member missing, in this case, it's Microsoft.

Ideally, users should control their own identity. The current thinking around identity is to work towards 'Open Identity': with users controlling their own profiles and storing their data at an independent place on the Web. Open identity is still at an early stage. However, from the perspective of Mobile Web 2.0, a simpler approach could suffice. **Let us not forget that we are making a phone call here!** A simpler interpretation of identity may suffice for our purposes. That identity could be **'reputation'** and this leads us to the idea of 'verified tags'.

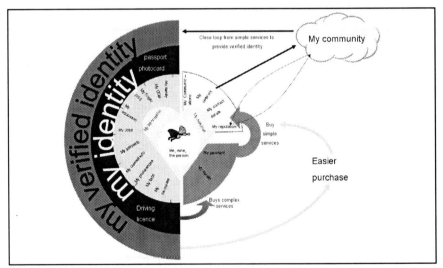

Figure 54: *Closed Loop Indentity Verification*

---

169 https://accountservices.passport.net/ppnetworkhome.srf?vv=400&lc=2057
170 http://www.projectliberty.org/

Ajit Jaokar and Tony Fish

Verified tags

As we have seen before, conferred identity (for example, a driving licence) can be used instead of the primary identity. For example, conferred identity can be used to buy a range of services in the physical world (for instance you can show your driving license to buy services). In the digital world, tags / avatars etc are all a form of identity. However, they are not verified.

To achieve the concept of 'I am not a number..' you need a verified tag.

How can we verify the tag? You could use something like the Liberty alliance or similar mechanism. But that's too 'top down', complex and expensive. But let us put this in perspective first; as indicated before, *A phone call is not a transaction!*. The stakes are a lot less lower. In this case, rather than a full fledged approach (such as liberty alliance), which is expensive, a simpler more organic approach could suffice. This approach is based on the concept of 'Identity = reputation'. **Reputation** is what others say about me on the Web. That's an organic way to 'verify' me i.e. my tag. A verified tag allows me to be identified and verified, but not because who I say I am, but because others can validate what I say. This makes the mechanism a 'closed loop' system. In contrast, the Internet allows me to set up an ID called 'John Smith', without proof. I can then take someone's verified ID and can communicate to another channel. As this is open; it can be abused.

The next logical question is: how to capture that 'reputation'? And here, lets introduce another word 'Pingerati . Trackbacks [171] are decentralised. In the sense, that when we trackback something, a link is created between the blog we are tracking back from and 'my' blog. Supposing, every time you tracked back, there was a 'ping' (hence Pingerati!) to another site; that site would then hold all your trackbacks ever.

Why is this useful?

Imagine there is a place on the Web where I could store all my trackbacks, then that place becomes an implicit profile. Maybe it could even be my own blog. Thus, my blog would contain all my blog entries, all the comments on my blogs

---

171 http://en.wikipedia.org/wiki/Trackback

but also all the comments I have made on other people's blogs in a separate section. An algorithm (PeopleRank) could then process all that information to create a ranking/probability indicating who I am.

Addressing

So, thus far we have seen how we could identify a tag based reputation. Our goal is to make a 'Internet originated phone call' to that tag now that we have reliably identified the tag. The question arises, how to 'place' this call from the Web? If this is a mobile phone, how can we bridge the 'fixed to mobile' gap? From a telecoms perspective, to place an IP call to someone on a mobile IP network, you need their IP address. That address may be mapped to any identifier (tag, avatar etc), but an IP address is required. We also need to know presence information: Is the 'tag' available to accept phone calls. On first impressions, the requirement of knowing the IP address in advance sounds very daunting. In fact, it may not be so as we show below.

We can think of two ways you can 'call' a phone.

   a) SkypeOut and

   b) Naked SIP

SkypeOut: The first and the simplest is to use an enhanced version of a mechanism like 'SkypeOut'[172]. SkypeOut has a profile and can already call any phone including a mobile phone. So, we need the Identity and presence features which are missing at the moment. Presence can also be provided by the user manually or through rules (Between 7 to 9 in the evening, call me on my home number).

Naked SIP: "Naked SIP" is SIP without IMS. Please refer to the section on 'Naked SIP' for an explanation of it's significance. Naked SIP provides plus third party applications an addressing mechanism to a phone which is independent of a Telecoms network provider.

---

172 http://www.skype.com/products/skypeout/

Ajit Jaokar and Tony Fish

Conclusions and Observations

From the above discussion, we see that:

a) I am not a number; I am a tag can be implemented at multiple levels; either as a simple 'SkypeOut' call or through naked SIP in the future.

b) Identity can be implemented organically through 'Pingerati'.

c) To some extent, we already use the Web to check identity manually. How many times have you 'Googled' someone to 'check out' who they are? That's an ID check!

d) To recap, it's more about convenience. When Amazon was first launched, many people thought it was about cheaper books. Today, we don't expect cheap books from Amazon – but we get 'something more' i.e. the choice. Similarly, with this concept – you get all the numbers in one place, you get presence information and you can call the numbers from the Web.

References: Marc Canter : Breaking the web wide open [173]The disruptive potential of Wi-Fi and WiMAX

IMS : IP Multimedia Subsystem

A discussion on the implementation and impact on IMS

Introduction

We had been following the hype around IMS (IP Multimedia Subsystem) with some scepticism. On first glance, it sounded depressingly familiar, trailing the ghosts of long dead previous hyped up technologies (WAP. MMS, LBS etc etc). Most of the current media interest focuses on the question 'Will IMS succeed?' As we shall see below, this is a silly question often asked by people who are not familiar with the big picture (typically people who approach the industry from the business/marketing side and don't understand technology)

---

173 http://www.alwayson-network.com/comments.php?id=12412_0_1_0_C

However, when VON pioneer Jeff Pulver[174], (whose views we respect), compared IMS to 'the return of the Sith' [175], we were intrigued. Thus, for the first time, we are talking not of the success (or otherwise) of IMS applications but of deeper (and darker?) motivations behind the hype surrounding IMS.

Of Sith and Sith lords

The return of the Sith is a Star Wars[176] reference: meaning an old empire clinging to power. When we wrote OpenGardens[177], in Dec 2004, the walled garden edifice of the Mobile Network Operators seemed unassailable. It was deemed to be the ideal business model i.e. the Mobile Network Operator should be both the carrier of data as well as the provider of value added services. The services provided by the operator are, in some way, preferred to third party services. Users did not have open access to the Internet.

In a little less than two years, everything has changed. Here are two examples:

a) The biggest exponent of the walled gardens model was the UK operator '3'[178], (others followed a milder version of the '3' mindset but '3' was the operator which blocked all Internet access (i.e. full walled garden)). Has the '3' strategy worked? As per timesonline[179] , Three UK says its ARPU is 34.50 UKP (50 Euro/60 USD). Of that, 25% is data. The majority of data is SMS. The point to note is: A majority of the data is **still** SMS, and that is no different from any other Mobile Network Operator. If the walled gardens strategy was working, then a majority of the data should be 'rich media' (songs, movie clips etc), not SMS.

b) Did you notice that it's a bad time to say that you are a mobile operator? The city is sceptical of the business model and the stocks of many of the Mobile Network Operators are stagnant.

---

174 http://www.jeffpulver.com
175 http://www.theregister.co.uk/2006/02/24/telcos_collaborate_for_ims/
176 http://www.starwars.com/
177 http://www.opengardens.futuretext.com
178 http://www.three.co.uk
179 http://www.timesonline.co.uk/newspaper/0,,2769-2189505,00.html

---

Ajit Jaokar and Tony Fish

In the UK, Orange was just the latest company to announce that they are 'not' a mobile operator (rather they are some kind of a new converged communications company[180]. That's a far cry from the strategy only a few years ago (voice revenues are going to decline but data (read some form of walled gardens) will rescue us)).

So, the dark forces of walled garden mindset are on the run...

But... is there a Sith Lord plotting a comeback to the walled gardens?

Is that walled garden 'IMS'?

What is IMS

Wikipedia provides a comprehensive definition of IMS181 .

> *The IP Multimedia Subsystem (IMS) is a standardised Next Generation Networking (NGN) architecture for telecom operators that want to provide mobile and fixed multimedia services. It uses a Voice-over-IP (VoIP) implementation based on a 3GPP standardised implementation of SIP, and runs over the standard Internet Protocol (IP). Existing phone systems (both packet-switched and circuit-switched) are supported.*

> *The aim of IMS is not only to provide new services but all the services, current and future, that the Internet provides. In this way, IMS will give network operators and service providers the ability to control and charge for each service. In addition, users have to be able to execute all their services when roaming as well as from their home networks. To achieve these goals, IMS uses open standard IP protocols, defined by the IETF. So, a multimedia session between two IMS users, between an IMS user and a user on the Internet, and between two users on the Internet is established using exactly the same protocol. Moreover, the interfaces for service developers are also based on IP protocols. This is why IMS truly merges the Internet with the Cellular world; it uses Cellular technologies to provide ubiquitous access and Internet technologies to provide appealing services.*

From the above definition, we see that SIP is at the heart of IMS.

---

180 http://news.zdnet.co.uk/communications/0,39020336,39274893,00.htm
181 http://en.wikipedia.org/wiki/IP_Multimedia_Subsystem

The following definition of SIP is adapted Wikipedia[182]

> *Session Initiation Protocol (SIP) is a protocol for initiating, modifying, and*
> *terminating an interactive user session that involves multimedia elements*
> *such as video, voice, instant messaging, online games, and virtual reality. It*
> *is one of the leading signalling protocols for Voice over IP, along with H.323.*
>
> *SIP clients traditionally use TCP and UDP port 5060 to connect to SIP*
> *servers and other SIP endpoints. SIP is primarily used in setting up*
> *and tearing down voice or video calls. However, it can be used in any*
> *application where session initiation is a requirement. A motivating goal*
> *for SIP was to provide a signalling and call setup protocol for IP-based*
> *communications that can support a superset of the call processing functions*
> *and features present in the public switched telephone network (PSTN). The*
> *SIP Protocol by itself does not define these features, rather, its focus is call-*
> *setup and signalling.*
>
> *Some basic functions being implemented by SIP are currently being done*
> *by the SS7 network (signalling system 7 network) within the telecoms*
> *industry. SIP can do what SS7 does currently but it can do a lot more*
> *especially when it comes to multimedia sessions.*
>
> *SIP is a simple peer-to-peer protocol. It is similar to HTTP and shares some*
> *of its design principles: for example it is human readable and request-*
> *response structured.*

IMS was originally designed for Mobile networks, but with release 7 of IMS, fixed networks are also supported. This led to Fixed/Mobile convergence – one of the main drivers for the industry today.

The motivation and inevitability of IMS

a) The business case underlying IMS: Inspite of all the hype and the controversy around IMS, we believe IMS will be implemented. There is a simple reason why. That's because, from a Mobile Network Operator standpoint, there is a clear business case for IMS. If internally (i.e. within the Operator's ecosystem), calls are handled using IP technology, then the Operator's cost base decreases (and hence profits increase). That's why we say it's a silly idea to question the possibility of IMS itself.

---

182 http://en.wikipedia.org/wiki/Session_Initiation_Protocol

b) The question mark around IMS services: The real question is, how will IMS impact services for the end user (and that includes new services which are yet to be created). That, we are much less optimistic about. An 'IMS' service would be a contradiction in terms since almost any service will be based on IMS. But certain services will be possible only through IMS. These will need SIP based handsets. It is debatable if users will want these handsets or if they will be available in tandem with the network infrastructure.

c) An analogy with 3G: We saw the same analogy with 3G. A lot was said about the viability and possibility of 3G. Today, 3G exists in most parts of the world. But the sad thing is, 3G services have yet to take off. Inspite of all the hype surrounding 3G, Operators had a business case for 3G because network capacity was already strained and needed to increase (and that includes network capacity to handle voice). So, 3G was a 'no lose' situation in any case for the Operator. IMS is the same because there is a defensive reason for Mobile Network Operators to deploy IMS (as per point (a) above).

d) The offensive reasons for deploying IMS: Include the much publicised new services such as push to talk, presence based services, gaming, multimedia conferencing etc. We are sceptical of their success.

e) Fixed to mobile convergence: Both fixed line and Mobile Operators are deploying IMS because they see growth in getting a share of each other's businesses i.e. Mobile Operators want to get some fixed line customers and vice versa.

f) The hype from other players: Besides the motivations of the Telecoms operators themselves, there are many vendors who are also bullish about IMS. Infrastructure vendors like Alcatel and Cisco see IMS as an opportunity to sell more equipment. Consultancies like Accenture and Cap Gemini see the potential of implementation fees. So, suddenly, we see a lot of activity in the industry stirred up by all these players.

g) Overcoming the threat of VoIP: One of the motivations behind IMS was to overcome the threat of VoIP. But VoIP is free! It's very hard to claim VoIP and then charge a cost (which is what telecoms vendors would have to do to cover their investment).

The dark side of IMS

On first impressions, IMS seems to open up the Telecoms architecture to third parties such as IT companies and other providers. The adoption of the IP protocol within the Telecoms network also appears to be a positive development.

But there is a problem...

We can best understand it by comparing it to a wider debate about net neutrality taking place in the industry today, with net stalwarts like Sir Tim Berners-Lee[183], weighing in their arguments in support of a neutral Internet (all packets should be treated equally).

The save the Internet site [184]elaborates this debate succinctly on its website:

What is saving the Internet all about?

This is about Internet freedom. "Network Neutrality" -- the First Amendment of the Internet -- ensures that the public can view the smallest blog just as easily as the largest corporate Web site by preventing Internet companies like AT&T from rigging the playing field for only the highest-paying sites. But Internet providers like AT&T, Verizon and Comcast are spending millions of dollars lobbying Congress to gut Net Neutrality. If Congress doesn't take action now to implement meaningful network neutrality provisions, the future of the Internet is at risk.

What is network neutrality?

Network Neutrality — or "Net Neutrality" for short — is the guiding principle that preserves the free and open Internet. Net Neutrality ensures that all users can access the content or run the applications and devices of their choice. With Net Neutrality, the network's only job is to move data

---

183 http://news.bbc.co.uk/1/hi/technology/5009250.stm
184 http://www.savetheinternet.com

Ajit Jaokar and Tony Fish

— not choose which data to privilege with higher quality service. Net Neutrality is the reason why the Internet has driven economic innovation, democratic participation, and free speech online. It's why the Internet has become an unrivalled environment for open communications, civic involvement and free speech.

IMS is the same effect played out on the Mobile Data Industry.

Hence, it is a potential walled garden at the 'packet' level. And that's why we compare it to Sith Lords! It's the final attempt to recreate a model which is already doomed. We are not the only ones who think so. Telecoms analyst Martin Geddes[185], says in an article published by the Moriana group[186]

It (IMS) attempts to capture the flexibility and ubiquity of Internet Protocol whist ditching much of the Internet's design philosophy.

That sentence captures the essence of the issue! The Internet succeeded precisely because it was designed as a dumb pipe with the intelligence concentrated around the periphery. In other words, intelligent nodes and dumb pipes go together. 'Dumb pipes' means: all packets are created equal. There is no intelligence in the network, only in the nodes.

IMS uses IP, BUT adds some intelligence in the network because it does NOT treat all packets equally. IMS promises to improve the quality of service, reduce SPAM, provide better quality rich applications (such as video) and so on. The often hidden caveat of this promise is the understanding that packets may not all be treated equally. In financial terms, it translates to premium prices for premium connectivity.

Conclusions

In conclusion, we believe

    a) Will IMS implementations succeed? That's like asking will 3G implementations succeed. Yes, they will.

---

185 http://www.telepocalypse.net
186 http://www.morianagroup.com

b) Will IMS services succeed? We are not too optimistic

c) Is IMS 'good' for the industry as a whole? In our view, anything that violates the principles of net neutrality aka 'Sith Lords' is not good for the industry. It all depends on the mindset of the Telecoms implementations. QOS (Quality of service) is a good thing, but tiered price points based on the same principles is not.

IMS: The significance of 'Naked SIP'

This section discusses the interrelationship and significance of IMS(IP Multimedia Subsystem) and SIP(Session Initiation Protocol)

"Naked SIP" is SIP without IMS. Naked SIP is disruptive, as we shall see below. Let us first recap the concepts of SIP and IMS. Wikipedia provides a comprehensive definition of IMS:

> The IP Multimedia Subsystem (IMS) is a standardised Next Generation Networking (NGN) architecture for telecom operators that want to provide mobile and fixed multimedia services. It uses a Voice-over-IP (VoIP) implementation based on a 3GPP standardised implementation of SIP, and runs over the standard Internet Protocol (IP). Existing phone systems (both packet-switched and circuit-switched) are supported.

> The aim of IMS is not only to provide new services but all the services, current and future, that the Internet provides. In this way, IMS will give network operators and service providers the ability to control and charge for each service. In addition, users have to be able to execute all their services when roaming as well as from their home networks. To achieve these goals, IMS uses open standard IP protocols, defined by the IETF. So, a multimedia session between two IMS users, between an IMS user and a user on the Internet, and between two users on the Internet is established using exactly the same protocol. Moreover, the interfaces for service developers are also based on IP protocols. This is why IMS truly merges the Internet with the Cellular world; it uses Cellular technologies to provide ubiquitous access and Internet technologies to provide appealing services.

From the above definition, we see that SIP is at the heart of IMS. The following definition of SIP is adapted Wikipedia:

Session Initiation Protocol (SIP) is a protocol for initiating, modifying, and terminating an interactive user session that involves multimedia elements such as video, voice, instant messaging, online games, and virtual reality. It is one of the leading signalling protocols for Voice over IP, along with H.323.

SIP clients traditionally use TCP and UDP port 5060 to connect to SIP servers and other SIP endpoints. SIP is primarily used in setting up and tearing down voice or video calls. However, it can be used in any application where session initiation is a requirement. A motivating goal for SIP was to provide a signalling and call setup protocol for IP-based communications that can support a superset of the call processing functions and features present in the public switched telephone network (PSTN). The SIP Protocol by itself does not define these features, rather, its focus is call-setup and signalling. Some basic functions being implemented by SIP are currently being done by the SS7 network (signalling system 7 network) within the telecoms industry. SIP can do what SS7 does currently but it can do a lot more especially when it comes to multimedia sessions.

From the above, we see that:

a) SIP is primarily involved in setting up and tearing down IP sessions.

b) On the Web, SIP is used for peer to peer sessions.

SIP and IMS go together because peer to peer does not work well with mobile. In a true peer to peer session, there would be no intervening server component. The devices would communicate directly with each other. This could lead to problems in some situations. For example, if the device is out of coverage, in the absence of a server component; no message could be left. Thus, a network component is required. That network component is IMS. Hence, in a traditional telecoms situation, IMS and SIP go together because SIP sets up and tears down the IP session and IMS provides the back end function including quality of service, support services to call management etc.

However, SIP and IMS need not necessarily go together!

A report published by Dean Bubley[187] who blogs at disruptive analysis blog[188] explores this topic further. In a nutshell, from the report highlights page:

- *SIP – an essential basic subcomponent of IMS – is much easier to implement than a full IMS software framework. SIP-capable phones are already shipping.*

- *There are many interesting non-IMS applications of SIP on mobile phones, such as VoIP, Internet IM, enterprise IP-PBX access, or interactive games.*

- *In total, 787m SIP-enabled mobile handsets will ship in 2011, of which 40% will be smartphones. Europe will account for 50%+ of SIP handset volume shipments until 2010, although Japan and Korea lead, in penetration terms.*

- *"Naked SIP" phones, on which 3rd-party applications can exploit the SIP stack, will grow rapidly in importance, with 48m shipping in 2006, more than 220m in 2008 and 500m+ in 2011. This is a huge threat to mobile operators.*

- *Naked SIP will be enabled by smartphones OS's, virtual machines like Java, and the inclusion of "exposed" SIP in many featurephone platforms.*

- *Although some devices will support both naked SIP and operator-oriented IMS applications, there will be 1 billion more naked SIP handsets shipped, than operator-only "closed IMS" phones, between 2006-2011.*

- *Internet brands, enterprises, 3rd-party developers and competing service providers will exploit the opportunities from the 1.6 billion "naked SIP" phones that will ship between now and 2011, using on-handset software clients.*

- *Some operators will attempt to block "parasitic" 3rd-party SIP applications, by "locking" handsets or intercepting network traffic. These attempts will seem clumsy and vindictive, and will likely drive churn and customer disloyalty.*

---

187 http://disruptivewireless.blogspot.com/
188 http://www.disruptive-analysis.com/sip_and_ims_handsets.htm

---

Ajit Jaokar and Tony Fish

So, what does this tell us?

'SIP only' phones could be numerous. They could run useful applications from third parties. They don't need IMS. They could be true 'peer to peer'. You could place a VoIP call to such phones from the Web because they contain the VoIP stack and the SIP stack. This makes (naked) SIP + third party applications and not IMS the more disruptive technology!

We use the concept of Naked SIP in the next section when we discuss the implementation of 'I am not a number, I am a tag'.

Principles: OpenGardens, Walled gardens and Mobile Web 2.0

A discussion on the impact of Open systems on Mobile Web 2.0

Open systems

Historically, major technology vendors have used the philosophy of "walled gardens", often with considerable success. However, over time, we see walled gardens crumbling all around us and being replaced by Open systems. If 'customer lock-in' was the byword of the older players, Web 2.0/Mobile Web 2.0 are all about relinquishing customer control. Note that, Openness is a philosophy as opposed to a specific technical standard. For instance, AOL, although deemed a walled garden, uses RSS feeds to open up their ecosystem. Similarly, Microsoft uses SIP (an open standard) for Internet telephony instead of using its own standard or another proprietary standard.

'Openness' itself can mean many things:

- Openness of access for the customer : The ability to access any content

- Open APIs/ Open platforms: A level playing field for third party applications as compared to the provider's applications and the ability for any developer to use a published API(Application programming interface) or

- 'Open source' as defined by http://www.opensource.org/docs/ definition_plain.php

There are many areas on the web where open systems are playing a role (and will play an even greater role in future). These include elements like Identity, Open Media, Microcontent Publishing, Tags and so on. What is apparent is: Open systems are fast becoming pervasive over the web. The Mobile ecosystem cannot remain in isolation locked up from the rest of the world for long.

OpenGardens and Walled gardens

We first coined the phrase 'OpenGardens' when we wrote the book late in 2004. OpenGardens is the opposite of walled gardens. A 'walled garden' is a mechanism to restrict the user to a defined environment i.e. forcing them by some means to stay within the confines of a digital space. This restriction, often defined by a single company, is a means of exercising control and supposedly maximising revenue.

What are the 'bricks in the wall' i.e. the elements that make up a walled garden?

A walled garden is any mechanism for an entity (not just a Mobile operator) to restrict the user experience by confining the user to a specific region / space as defined by the entity. The rationale supposedly is: the user is served better and the service is more profitable for the provider. In an Internet/Mobile environment, this can often take the guise of restricted browsing but it has other facets as we see below. From a developer perspective, a walled garden could mean 'restricted access', i.e. - your application in some way cannot access all customers OR the provider's application has access to some features that you cannot access. These restrictions can be commercial or technical. In conversational terms, walled gardens are deemed to mean any restriction placed on users or applications, which are aimed at confining the user to a set of features controlled by the provider.

Within the Mobile Data Industry, a Mobile operator has some elements that lend considerable power i.e. a large customer base, Knowledge of the subscriber's location, Billing relationship to the customer and Customer services and marketing reach. There are others especially on the voice side but these

Ajit Jaokar and Tony Fish

elements are critical for data applications. In addition, in a portal situation, the Mobile operator has the ability to control the positioning of the user on the menu, which is yet another 'Brick in the wall'. Extending the concept even further – Mobile Operators are not the only ones in the walled gardens game. Brands are often viewed as a safe bet - especially branded content. A friend uses the colourful and insightful expression 'Elephants mate with elephants'. This means the large content providers of the world may well have done deals with the large Mobile Operators leaving little scope for the smaller player. There is already evidence of this with some content deals in Europe.

The issue of walled gardens first arose with WAP (Wireless access protocol) phones, which are used to "browse" content. It arose due to a specific legal situation with a European Mobile operator who prevented users from changing the default settings on their phone. This means, users always started with a specific WAP site (i.e. home page) as directed by the Mobile operator and further they could not change the home page itself. While this model was commercially appealing to the Mobile operator and also the advertisers, it was not conducive to the small developer. A developer successfully appealed against the Mobile operator and won. In retrospect, the whole issue seems irrelevant in the case of WAP - because for various reasons, consumers never used WAP sites in large numbers.

So, do walled gardens exist?

While the 'hardcoded WAP home page' does not, there are indeed other ways to create restrictions. A true OpenGardens ecosystem would exist if 'all applications had a level playing field'. Where 'menu positioning' seems to be the most obvious 'choke point', there are other ways to cripple applications belonging to external developers specifically if they are denied equal access to certain resources for example location information.

Certainly, Mobile Operators and developers have a different mindset. One person (working for a Mobile Operator) referred to a quaint phrase 'revenue leak' – and no prizes for guessing who they think the revenue is 'leaking to'. While most Mobile Operators genuinely don't know how to tackle the brave new world of applications development, there are other factors at play here. Some Mobile Operators are concerned about pricing a service 'wrongly' in their view. Their fear is – the service will be popular but not profitable. Hence, a further reluctance to relinquish control since a global 'free for all' access tends to reduce prices. The sad by-product of this limited thinking is: the service is never launched at all!

In addition, there are some issues which are controlled by neither the Mobile operator nor the developer: such as legal and statutory guidelines. Thanks to a certain measure of control, the Mobile Data Industry has been spared the worst excesses of the Internet such as SPAM, copyright violation, privacy violation etc. As the industry matures and adopts more guidelines: the question is that of 'intent'. Is control a sign of maturity or is it a smokescreen? Clearly, the Mobile operator cannot flow against the rising tide of the Internet and the global flow of information i.e. the Internet is the driver to mobility and not the other way round.

OpenGardens

The phrase 'OpenGardens' evokes a variety of responses, ranging from the open source evangelist who gets misty-eyed thinking of 'Linux on every Mobile device' to the Mobile operator who insists: 'There are no walled gardens!' OpenGardens is not an anti Mobile operator stance. But that does not mitigate the fact that the Mobile Network Operators are the biggest culprits. OpenGardens/Open systems primarily means 'open platforms' when applied to the Mobile Network Operator. By 'openness of platforms', in the Mobile applications context, we mean the Mobile operator's infrastructure. An OpenGardens ecosystem is a holistic/inclusive environment, which could foster the creation of next generation utility led (as opposed to existing entertainment led) Mobile applications. These applications could often span multiple technologies/concepts and are created by 'assembling together' a number of existing applications. In Web 2.0 terminology, that's a 'Mashup'.

Technologically, in its ultimate form, this approach can be viewed as 'API enabling' a Telecoms network. API (Applications Programming Interface) is the software that enables service provision by the Mobile operator. The external application can make a software call via the published API, thereby creating a 'plug and play' ecosystem. The API model is also called by other names such as 'networked model', 'Bazaar model' or 'web services model'.

Ajit Jaokar and Tony Fish

Mobile Web 2.0 and OpenGardens

It is in this context that we need to look at OpenGardens again. Specifically, as Web 2.0 and Mobile Web 2.0 become prevalent because Web 2.0 is all about harnessing collective intelligence. To harness collective intelligence effectively, we need a large 'body of intelligence' i.e. a large body of people. Walled gardens, closed standards, proprietary systems hinder Web 2.0 / collective intelligence in two ways: firstly, by reducing the number of participants in an ecosystem and secondly by limiting the activity of the participants who are in an ecosystem. The Web is one big ecosystem. By fragmenting the Web through proprietary standards, we create smaller silos within the one large ecosystem.

Will walled gardens work?

Will walled gardens work? Let's come back to AOL.

Walled gardens are not new. One of the best-known instances was the early AOL. On an extreme case, in the early days, users could not email others outside AOL! (remember this was only about eight years ago). However, the early users liked these restrictions since there was the perception of 'the big bad world out there' and AOL was deemed to be a trusted provider. As users matured, they realised that the restrictions were often a hindrance and ultimately, there are now a lot less restrictions on AOL.

It appears that walled gardens do not stand the test of time and become irrelevant as the medium matures. We believe this will happen in the Mobile Data Industry.

Principles: Unlimited use price plans for mobile data

Unlimited use price plans for mobile data could provide a major boost to applications like Mobile multiplayer games and Mobile Voice over IP.

'However, for mobile data, most markets don't currently have any (viable) unlimited use price plans. Nevertheless, both European and North American operators are taking steps to introduce unlimited use data price plans. At the moment these are mostly business oriented with limitations in use (for example - VoIP is restricted) although we are seeing development of the first consumer unlimited use price plans

Why have unlimited use price plans for mobile data?

Unlimited use price plans enable many applications which are currently not feasible. Take mobile multiplayer games. Single player mobile games are popular. Multiplayer games, in general, are hugely popular on other platforms (like PC and other games consoles, e.g. Xbox & PS2). Thus, one would expect that Mobile multiplayer games would be popular. However, despite early promise, Mobile multiplayer games haven't flourished throughout the world.

Similarly, VoIP (Voice over IP) is a popular service. By extension, one would expect *Mobile VoIP* to also be popular.

However, currently, mobile VoIP is only synonymous with 'voice over WiFi' i.e. voice calls made over a WiFi network. Although the WiFi network is growing fast, the world is far from being a 100% WiFi enabled space. This means, mobile VoIP suffers from the physical limitations of being near a WiFi hotspot.

The reasons that hamper both, Mobile Voice over IP and Mobile multiplayer games, is the lack of viable unlimited use data price plans. Both these applications are mobile bandwidth intensive and mobile data, by definition, is not free. Note that we are not advocating mobile data should be free: but rather that it should be affordable and more importantly, predictable in terms of costs. The best way to achieve this goal would be to develop unlimited use data price plans. As the name implies, 'unlimited use data price plans' are 'all you can use' data price plans for a fixed monthly data charge.

Admittedly, unlimited use price plans have been around for a while, but the catch lies in how 'affordable' they are, i.e. how high the barrier is set. Just a few months ago, they were as high as £75/month (140 USD), with restrictions on data usage along with restrictions on certain types of applications (such as music streaming, VoIP etc). In Europe and North America, with the launch of the T-mobile Web'n'Walk price plans, things may be changing. While these price plans are initially pitched to the business customer, some non-business price plans were also launched at the same time.

The Web'n'Walk Professional tariff was launched towards the business customer through a data card for laptops. It costs £19.99/month, plus the cost of the data cards. It also includes free WiFi hotspot access for twelve months.

These prices are reduced from the previous, artificially high rate of £45/month, a rate designed to restrict excessive data usage. With Web'n'Walk Professional, bandwidth intensive applications like VoIP are not permitted. In fact, there is no voice capability at all i.e. it's a 'data only' card. A more interesting trend lies in the consumer segment. For consumers, the T-mobile tariffs are as shown below:[189]

Price plan	Monthly charge	Inclusive Minutes	Inclusive Texts
Relax 20 + web 'n' walk	£27.50	100	25
Relax 25 + web 'n' walk	£32.50	150	100
Relax 30 + web 'n' walk	£37.50	250	100
Relax 35 + web 'n' walk	£42.50	500	200
Relax 50 + web 'n' walk	£57.50	750	200
Relax 75 + web 'n' walk	£82.50	1200	200

From the above table, we can see that the price plan itself is more complex and

---

189 http://www.t-mobile.co.uk/Dispatcher?menuid=plans_webnwalk_relaxplans&chosenPricePlan=pmt

expensive. Although there is unlimited surfing (with the usual restrictions on VoIP, etc), there is also the tempting (yet complicated) compensation of more inclusive minutes and texts. This means that browsing may still be hampered because the combined cost is still high.

We can compare this to a similar scenario in Japan.

In Jan 2006, NTT DoCoMo announced that the flat rate price plan of 4095 Yen ($36/month) would be extended to more popular voice packages[190]. Indeed, the 4095 yen per month price plan existed before Jan 2006, but the users were forced to combine it with a much more expensive voice plan, making the combined package very expensive.

We see a similar situation with the T-mobile announcement. Nevertheless, it is a step in the right direction and competitive positions may force other operators to respond with deals that are more favourable.

A more interesting situation arises with the restrictions on VoIP. There are precedents for Mobile Operators rolling out Skype. For example, in Germany, the operator E-Plus offers a data tariff with a Skype client. Indeed, T-mobile itself may introduce a VoIP client[191]. Therefore, true Mobile VoIP may well be on the cards soon.

We see a similar situation with Mobile multiplayer games. In Japan, it used to cost $22.18 to send one megabyte of data over the Vodafone network. This high cost meant that mobile multiplayer games hadn't flourished in the Japanese market. (Reference: Matthew Bellows says on Mobenta/WGR)

Contrast Japan with South Korea. Per megabyte fees for data transmission are approximately $2.69 on LG Telecom and $4.49 with KTF. Even more interestingly, according to Com2Us[192], operators share as much as 60% of the data revenue associated with multiplayer mobile games with the game developer. As you would expect, mobile multiplayer games, from traditional card games to massive multiplayer games are booming in Korea.[193]

---

190 http://www.wirelesswatch.jp/modules.php?name=News&file=article&sid=1745
191 http://news.zdnet.co.uk/communications/3ggprs/0,39020339,39267682,00.htm
192 www.com2us.com
193 http://opengardensblog.futuretext.com/archives/2005/06/packet_fees_boo.html

---

Ajit Jaokar and Tony Fish

Conclusion

Unlimited use price plans are vital for the uptake of mobile data applications. Currently, we don't see many instances of unlimited use price plans in Europe and in many parts of the world. We expect this to change in the very near future.

Principles: The disruptive potential of Wi-Fi and WiMAX

Can Wi-Fi and/or WiMAX be used to create an alternate Telecoms network?

Introduction

Many of us work with Mobile Network Operators and find that a frustrating experience (and that includes many people who work for Operators!). Secretly, we think that we could do a better job! Could we perhaps set up an alternative wireless network? The two most likely technologies that can achieve this goal are Wi-Fi and WiMAX.

Is this a pipe (no pun intended) dream? Can it be done?

We are considering two questions here:

a) Can alternative technologies like Wi-Fi and WiMAX be used to set up a functionally analogous Mobile Network Operator? And...

b) How disruptive can such an operation be? Who would it benefit?

Before we start: many of us, including me in this article, are guilty of combining 'Wi-Fi' and 'WiMAX' in the same argument. Other than similar sounding names, they are really quite different. Indeed, as we shall see below, WiMAX has more in common with 3G (i.e. Cellular) and broadband than it does with Wi-Fi. So, our roadmap for this section will be:

a) Understanding the evolution of Wi-Fi to Wi-Fi grid networks and the possibility of using Wi-Fi grid networks as an operator.

b) The potential of WiMAX technology to provide an alternate wireless network.

But, to really understand this issue, we have to consider the question:

1. 'What is 'mobile'? And a secondary question…

2. 'What do we mean by a 'functionally analogous' Mobile Network Operator?'

When we refer to mobility, we are referring to 'anywhere/anytime' access to the service. The caveats to this definition are:

a) A reasonable quality of service is expected.

b) The lack of quality of service could lead to a 'good enough' solution.

c) A good enough solution, if it is cheap, could actually be a preferred solution in comparison to a complex, expensive, non interoperable solution.

d) In some cases, 'anywhere / anytime' could equal 'sometimes'. Specifically, 'sometimes' equates to proximity to a hotspot. Thus, the requirement is really a trade-off between quality, coverage, cost etc.

Another way to put this is: 3G and Wi-Fi don't compete at one level because 3G is a Cellular technology (in theory, available any time) and Wi-Fi works only around a hotspot. But, a 'Wi-Fi Blackberry' is already here! The customer will spend their dollars at only one place. So, at a dollar level, they compete. This comparison will be played out on a much larger scale soon with the introduction of WiMAX. In the end, the customer will decide the scenario that will succeed.

With this background, let us consider the two scenarios: namely Wi-Fi mesh networks and WiMAX.

Wi-Fi mesh networks

Wi-Fi or Wireless LANs is a term which refers to a set of products that are based on the IEEE 802.11 specifications. The most popular and widely used Wireless LAN standard at the moment is 802.11b, which operates in the 2.4GHz spectrum along with cordless phones, microwave ovens and Bluetooth. Wi-Fi enabled computers or PDA (personal digital assistants) can connect to the Internet when in the proximity of an access point popularly called a 'hotspot'. The Wi-Fi alliance is responsible for governing and standardising Wi-Fi technology.

Due to the (runaway) success of Wi-Fi, the industry has always looked for ways to extend the role of Wi-Fi beyond simple localised access. Most people are familiar with the 'Basic' configuration of a Wi-Fi network. In this case, the Wi-Fi access point is directly connected to a network and provides wireless access around a 100m radius around the access point. The basic configuration is not scaleable. An alternative configuration is called a **'Wi-Fi mesh'** configuration. In a Wi-Fi mesh configuration, only one access point needs to be connected to a physical network. Each access point in a mesh network supports two connection modes: A client connection, through which it communicates with a client which wants to connect to the network and a backbone connection, which it uses to connect to the client in the network which is connected to the physical network

Thus, Wi-Fi mesh technology offers a way to achieve, in principle, coverage around the whole metropolitan area.

WiMAX

WiMAX stands for wireless interoperability for Microwave access and is defined by IEEE 802.16 standards. It is promoted and standardised by the WiMAX forum. The confusion around WiMAX arises because there are two versions of WiMAX: fixed location WiMAX and Mobile WiMAX. The original specification of WiMAX was for the fixed location WiMAX (IEEE 802.16-2004). Fixed location WiMAX is mainly used to provide wireless broadband connectivity. It may be ideally suited in many rural areas but also in some urban areas where installation of physical connections is expensive.

Mobile WiMAX (specified in 802.16e-2005), on the other hand, evolved much later (and in fact, is still evolving). In contrast, fixed location WiMAX is relatively mature with the first products hitting the marketplace already. Unlike fixed location WiMAX, Mobile WiMAX competes with Cellular services. Mobile WiMAX relates to 'handoff' i.e. a customer can move from one base station to another and yet maintain their network connection. As you can gather, Mobile WiMAX competes with both 3G Cellular services and also with the Wi-Fi mesh networks.

Currently, much of the WiMAX activity is around fixed location WiMAX (for instance clearwire[194] recently got a close to a billion dollars of funding!). And much of the hype is around Mobile WiMAX. In between these two modes, there is the nomadic WiMAX mode. Nomadic WiMAX means you have coverage within the area but not handoff (so the connection drops when you go outside the coverage area). Considering the relatively wide coverage area, this may not be so bad.

Spectrum issues

Spectrum issues will affect the actual possibilities and rollout timeframes of these technologies. They differ from country to country. Wi-Fi operates in the unlicensed spectrum with 802.11b in the 2.4 GHz and 802.11a in the 5Ghz. WiMAX on the other hand can operate in either licensed or unlicensed frequencies from 2 GHz to 66 GHz. The original versions of WiMAX operated in the 10-66Ghz range but needed 'line of sight' access between the base station and the client. Obviously, this was very restrictive. Subsequent WiMAX standards are designed to operate in the lower frequencies (2 GHz to 6 GHz). At these frequencies, line of sight is not required.

Observations

a) Many people consider Wi-Fi, as it stands today, to be disruptive. Certainly, it is. However, the future appears far more disruptive especially when we consider WiMAX and Wi-Fi mesh technologies.

---

194 www.clearwire.com

Ajit Jaokar and Tony Fish

b) Simple always wins; every time.

c) The industry uses popular technologies in ways that were not originally envisaged (PLSQL instead of SQL, Ajax frameworks and Wi-Fi mesh networks are all examples of this trend.)

d) A good enough solution may be the winner rather than a complex, difficult and expensive solution.

Conclusions

We started off with the two questions:

a) Can alternative technologies like Wi-Fi and WiMAX be used to set up a functionally analogous Mobile Network Operator? And

b) How disruptive can such an operation be? Who would it benefit?

The answer is: It depends. Depends on what you call a 'mobile network'. More specifically, depending on what customers call a mobile network. Where will customers see value? As we have seen above, will customers accept trade-offs between handoff, QOS, Cost and coverage?

The meaning and definition of a 'network' in the customers' mind is changing! And that change means: a functionally analogous network (which will imply a trade-off between handoff, QOS, cost, complexity, coverage and interoperability) could well be built using Wi-Fi mesh and / or WiMAX

Commercially, it also comes down to spectrum issues. Who develops the spectrum and what services do they provide? One thing is for sure: setting up a Wi-Fi mesh network or a WiMAX network is a non trivial exercise. The companies who are likely to succeed in this task are the ones which are very well funded. They are also likely to be non-incumbents.

In or view, the winner will be the customer and the customer will define and decide what exactly they mean by a mobile network and whether it will succeed.

References

a) **WiMAX and the Metro wireless market: WiMAX vs. Wi-Fi and 3G**
Michael F. Finneran dBrn Associates, Inc. 189 Curtis Road Woodmere, NY 11598
Tel. (516) 569-4557 Mobile (516) 410-5217 mfinneran@att.net

b) **WiMAX: Opportunity or Hype?**
Michael Richardson
Patrick Ryan
University of Colorado at Boulder
Interdisciplinary Telecommunications Program

PART THREE

Funding, Business Models And Strategy

Introduction

We finally come to the third and the last part of this book. Part three pertains to funding, commercials and business models. It's scope is wider than Mobile Web 2.0 and it covers both Web 2.0 and funding in some detail.

Getting your venture funded

This section talks about funding new mobile related ventures. Funding is a critical component in the success of a new venture; hence, it is one of the factors that influence the outcome of Mobile Web 2.0 companies. However, the emphasis of the chapter is somewhat generic and provides a 'bird's eye view' on the status of the funding for mobile technologies from the US, EU and Asia. In the most simple terms getting a venture funded is about a team telling a story; honestly!

If a business plan is a story, then there are two considerations. First, when should the story start? Second, how much detail should the story include?

To understand these two considerations; here is a simple analogy about telling the story of your life. Should your story start at the point when the parents met? At your birth or after your first day at work? ; since they all help to set the landscape and context. From all starting points the story will eventually end up at some position. Having found this position there is a need to decide about the level of detail that is going to be used to explain the life story. Should it detail every birthday, every heart breaking scenario or every friend met?

Far too many business plans are presented that are unable to tell the potential investor what the business is actually about, what business it will be in, what the objective of the business is, and why the founders have a vision.

A good business plan will detail a coherent story. It will not spend too long setting a scene with no foundation or start point. Poor plans explain in too much detail, too many bumps and bruises and forget to explain how the bumps and bruises were actually gained (the learning).

Venture Funding would better be described as Venture "fundology". Funding is an 'ology; it is not an exact science. It has no absolutes (other than the limits set by legal best practice), but nevertheless, the process follows some guiding principles.

This chapter is set out into five sections.

- The first section is about the status of the US, EU and Asia funding markets.

- The second section is about preparing the plan,

- The third section is about the funding process, types, and terms.

- The fourth section is about numbers and

- The final, fifth, section is about managing your new shareholders and the new problems this brings.

Status of the financial market

Change is the only constant in the financial markets. On a daily basis, the value of a publicly quoted company is affected by many factors such as market sentiment to the sector the company is in, results of other companies in that sector, legal and regulatory changes, consumer demand, political factors and so on. For the most part, the management of a company cannot control these factors but it can influence them.

2006 Outlook and attitude towards mobile and wireless in Europe.

Observations

In 2006, the European mobile market is forecast to reach a stage of maturity, which is reflected by lower growth in subscriber figures as penetration levels of 80% to 100% are already present in a number of countries and there is a lower rate of introduction of new services and applications. At the same time, the pressure on prices by the service providers continues to increase fuelled by the further upcoming of MVNOs, branded resellers and cheap operator brands.

2005 saw a more widespread launch of new brands in the market in the major European countries, such as debitel in France, debitel light, Simyo and Base in Germany and DotMobile and easyMobile in the UK. This trend is likely to continue in 2006 with TV stations, newspapers and clubs looking to enter the market. In order to pre-empt the MVNO successes of the Nordic countries and the UK, the German carriers and service providers launched all their own low price brands, such as E-Plus' Simyo and Base brands.

The Mobile applications market for ringtones, wallpapers and games shifted significantly from print to TV advertising during 2005. Basically, TV channels have experienced a major impact through the link up with the mobile industry. While mobile portals such as Verisign's Jamba have continued to dominate advertising on youth oriented channels, all TV stations have adopted some form of interactivity through SMS and IVR based systems. However, consumer activists had some limited success in protecting the youngest mobile users from entering subscription-based relationships, which are highly profitable for the providers.

The premium SMS and connectivity market will continue to show little sign of growth as pricing pressures continue to exist with a couple of dozen providers in nearly every country. With low barriers to entry, the market will see further providers entering the market and consolidation of SMS connectivity and application providers. It is forecast by GSMA that during 2007/8 there will be an acceleration in the number of absolute UMTS subscribers with the uptake driven by price and a strong marketing push by the network operators, rather than by new innovative services, such as video streaming, MP3 downloads and video conferencing, which continue to be marginal revenue generators because of the lack of widespread availability of suitable handsets and the cumbersome operation of the handset set-ups. It is unlikely that many of these services will become mass-market in Europe until the suitable handsets are also in the hands of the likely users, i.e. the youth market.

The consolidation of the European Mobile applications market will be largely completed by year end 2006 with fewer companies available for M&A. In 2006/7, it is expected that a number of larger application players will be acquired by the large US-based global media companies and portals as well as Japanese mobile application providers. However, 2006/7 will see numerous transactions in the mobile middleware and software space as the consolidation in that market has only just started (Qtr 2 2006).

Whilst the European Mobile/Cellular sector has reached maturity, in many aspects, it still provides lots of space for truly innovative solutions, applications and technologies. Given the high subscriber penetration and the increasing competition from MVNOs on basic voice pricing, the mobile carriers need to refocus during 2006/7 on the differentiation on new services and applications if they do want to give up their pricing levels.

Many mobile software and application companies that were funded in the late 90s and early 2000 are looking back on a number of tough years with limited business success. The survivors of the big shake-out over the last years are now looking frequently for some kind of M&A activity or even an IPO.

Hot topics

- Mobile gambling – driven by the highly successful 2005, with IPOs of Internet gambling companies on the London Stock Exchange raising millions of pounds, the sector has received tremendous attention. Thus, the next big wave is expected for gambling on mobile devices.

- IMS - Multimedia services is a large growth area in the future development of the mobile industry and will change the overall structure of the technology platforms installed. 2006 will highlight an increasing number of companies who have developed a value proposition in the sector. The venture capital community is already looking at this sector with great interest.

- Convergence/Triple (quad) Play – The market is moving finally towards full convergence with triple play to become more widely spread. As the market is still early, it is open for any new technology solutions with a clear-cut role in the market. Evidence of this could be seen through the NTL/ VirginMobile merger or the acquisition of EasyNet by Sky.

- VoIP – Several start-ups including Skype are working to bring VoIP to the mobile world. Given the high prices of mobile in comparison to fixed voice the potential market is very large and entirely unexploited. The larger number dual-mode WiFi/GSM handsets will enable mobile VoIP to gain market share during 2006.

- Metadata – There is an increased awareness that metadata in all forms will provide a unique database and descriptor of users. Of most interest is how to capture this data without user intervention but with prior knowledge, and the subsequent analysis to create an improvement.

What do founders and investors need/look for in Europe?

The European market for venture capital for the Mobile/Cellular industry has been a lot more limited than the US and it is likely that capital is going to be less freely available during 2006. Actually, in all European countries (with the exception of the UK) there are only a handful (or maximum a dozen) of venture capital funds which are able to invest in new wireless companies.

During 2006, because of the limited fundraising opportunities from traditional European venture capital sources, founders and investors need to look at alternative ways of raising additional funding early in their company lifetime. Main sources for additional funding are:

1) The US venture capital market, usually by flipping the company to a US corporation; and 2) A relatively small scale IPO on the London AIM segment or any of the other exchanges across the continent (e.g. Swiss or Munich Stock Exchange.)

The markets will always like:-

- Unique technology advantage and the ability to sustain it. European investors are generally very technology focused and are expecting a strong intellectual property position ideally built around defendable patents.

- Global market opportunity with the ability to build an EUR 100 million company at the time of exit. Most VCs have become very selective and will continue to be so during 2006. Thus, they are focusing their efforts only on companies that can address a global market opportunity and have the potential to obtain an exit valuation of at least EUR 100 million.

- Experienced team. A team with relevant background and expertise is always crucial for the success of any start-up company, both in raising funds and in being commercially successful. Repeat entrepreneurs are rarer in Europe than they are in the US. Thus, being able to recruit the right talent is more difficult.

2006 Outlook and attitude towards mobile and wireless in the United States

Observations

The US market is entering its second year of premium billing (2006), exponential growth in the number of mobile & wireless services, and increased mobile commerce revenues. End user uptake in devices, premium SMS service offerings, as well as an increasing number of mass-media advertising partnerships will also serve as growth stimulus for the market.

The media industry's continued adoption of mobile distribution channels also indicates growing support for cross-media promotional services and campaign

Ajit Jaokar and Tony Fish

installations. Where branded mobile storefront installations dominated business revenues in 2005, companies like m-Qube are increasingly moving into a dominant position in this saturated business segment. Mobile content and application distribution will continue to grow, following a pattern similar to mobile application service provider markets in Western Europe.

From a financial market point-of-view, mergers and acquisitions within the mobile content sector will continue, appealing mobile content offerings, including those filling a niche market, remain attractive for M&A. But whereas advertising investment previously defined the dominant strategy, focus has shifted towards media advertising partnerships and revenue-share business models for larger mass-media companies. As content consolidations have centralised the access of mobile content, and the launch of branded mobile content portals requires a large marketing investment, mobile service companies with mass-media partnership strategies are increasingly attracting the attention of the larger, international media companies.

Over the past two years, MTV has led the way by acquiring the rights for an in-house mobile service platform and the US-based MVNO, AMP'ed, made a similar move, highlighting the trend towards an in-house mobile campaign and content management program, and then MTV invested in them. A data-rich multimedia platform with a content management system, strong CRM and data mining capabilities, as well as traffic and billing support for rich media services like video and audio streaming, will enjoy larger success in the United States, compared to most European markets. All major Mobile Operators have effectively and aggressively moved towards video, TV and full-length MP3 offerings tailoring content for use on new handsets and new networks.

Hot topics

- Streaming Media – As an intensely media-centric market, US consumers might adopt Mobile TV and on-demand services such as music and video more strongly than European equivalents. Businesses targeting media companies as their core-customer, circumventing Mobile Operator dependence and relying on media for promotion, will enjoy strong market traction over the next 12 months.

- Mobile text, photo and video blogging – Recent explosive growth by companies like Myspace.com and others, brings about a powerful, American-born technology trend: blogging. Millions and millions of Americans have their own blog sites, personal or professional. In addition, the mobile inbox notion of letting users direct traffic into personal online environments from their phone will also find commercial traction over the next 12 months.

- RSS (Real Simple Syndication) – Mostly considered a grassroots technology where users can (and do) organise news and content feeds into one domain are currently a must-have on most news sites. Where technology-savvy analysts and journalists use blogging to share personal opinion, RSS represents the gathering and publication of relevant data. Where RSS is emerging on the Web, its Mobile applications have yet to trickle out to the commercial market place.

- Mobile Flash – As handset screen technologies and network bandwidth are increasing in capacity, and as Macromedia's Flash moves further into the installed base of mobile handsets around the world, the mobile industry will experience a fundamental change. From being a decoration-centric individual entertainment experience focused on personal play, emerging consumer demand will generate mobile offerings combining successes such as mobile community and game play with mobile content commerce and rich media streaming to create mobile companion-like portals. Whereas delivery and billing is still managed via WAP Push and Premium SMS, mobile flash technologies bring about a whole new design and experience capability, which J2ME failed to deliver. As flash can reside on the phone and easily connect to online synchronisation services, users can easily stay entertained and exposed to premium services for longer periods of time.

What stage are companies at, in the wireless mobile space.

While still generally perceived as a market in its infancy, the US market has leaped ahead with Mobile TV and Streaming services. As media companies force Mobile Operators to continuously keep innovating, and capital rich investors keep forcing companies to think big, the US market still lacks the experience to build "Bread and Butter" services, that over time will earn sizeable revenues. The expectation to grow at impossible speed will force many of today's market leaders to seek real-world experience in service packaging and design. As MMS and ring-back tones are pioneering technologies, premium SMS driven polyphonic ring tones and wallpaper "stores" that are proven business models in Europe with sizeable revenues, the US market has yet to understand the potential of print and TV advertising.

Mature technology offerings enabling broad data mining and service customisation in combination with proven traffic capacity and real life service deployment experience remain valuable selling points, bringing new companies on par with local, better-known competing companies.

What do founders and investors need/look for?

The team and its track record remain on top of the agenda. A business team has to prove deep industry knowledge, as well as a real ability to drive customer growth. As national TV is incorporating "premium" voting for its reality shows (Big Brother 6, Americas Choice 08/05), platform providers offering total "white label" solutions for media companies will increasingly attract the attention of the investment community. Companies that combine contributions of the telecom and media industries will be of great interest as the TV, Music and Movie industries look towards mobile solutions to generate new revenues on existing content.

Long-term technology strategies are needed to show direction in a rapidly changing market place. With so much money at stake, a company's ability to innovate and foresee emerging markets will attract larger M&A investors, producing a global strategy. Many commentators have observed the US market is trailing the European market by 16-18 months. As market conditions are rapidly changing with the incremental resolution of network compatibility issues, the European advantage still remains. What had been 16-18 months ahead just one year ago has now seen development times cut in half to perhaps nine months today (2006).

2006 Outlook and attitude towards mobile and wireless in Asia Pacific

Observations

While the US market can be defined as a single market for mobile and wireless services, Asia Pacific, somewhat like Europe, but without a common-law or currency, needs to be assessed on a country-by-country basis given the very different character of each individual market. In particular the level of maturity of ICT services in each regional territory, which defines the possibilities for certain trends to appear and accelerate at a certain point in time.

However, in general, Asia Pacific is driven by the world's great expectations on China and India. Both markets dwarf the growth prospects of any other Asian market due to sheer potential, their recent emergence as ICT leaders and dazzling speed of customer uptake in each respective domestic market (China alone today represents in excess of 20% of Asia's total telecom market and is still growing at an unparalleled pace). More traditional technology markets like Japan, Korea and Hong Kong have mature mobile and wireless sectors which drive global trends given their innovative nature in consumer buying behaviour and the existence of some of the world's leading technology players like Sony, Samsung and NTT DoCoMo.

Growth rates for fixed line, mobile and Internet users over the last years have soared in several countries across Asia. Mobile subscribers skyrocketed by 31% per year between 2000-2003 to touch 560 million — overtaking North America as the world's largest market. Broadband penetration in Asia is the highest in the world.

Asia-Pacific has a mix of mature and emerging markets, so total regional revenue growth is more muted, although there are very promising country opportunities within the region. Six of the nine fastest growing international markets are located within Asia. These emerging markets have truly unique characteristics, resulting in strongly differing mobile subscriber behaviour. Not only will consumer behaviour patterns vary, but subscribers will also have low budgets and be low-frequency users, which again indicates lower potential

Ajit Jaokar and Tony Fish

revenues. Operators and vendors' business models have been modified across emerging markets to allow for this. Pricing plans and handsets need to be offered more economically, but are structured in such ways that they do not cannibalise revenues from existing high-value users.

Key trends in the Asian wireless space have been identified:

- Keywords are education and awareness, affordability, community based initiatives and locally relevant content development, and more technology specific, the new digital identity, security, consumer behaviour of today's youth and converging technologies;

- Asian mobile commerce has reached around $18 billion in revenue in 2004 (a 40% market share) with around 1 billion mobile phone users, at the expense of yields per customer which dampens long-term revenue growth;

- Asia leads the global market for 3G revenues driven by some 120 million subscribers, whereby Japan and Korea take the helm at CDMA and W-CDMA. Their new digital media band satellite will beam 40 Korean and 70 Japanese TV channels to mobile subscribers;

- The majority of Asians are expected to use handsets as the primary mode for accessing e-mail and news with most subscribers using them for payment gateways;

- More than half of the new telephony lines installed in Asia will be pure IP rather than IP enabled, partially driven by the need to replace an ageing installed PABX base;

- A fundamental change is appearing in the way we think about streaming content, through less emphasis on video-on-demand and more emphasis on one-to-many broadcasting via satellite to portable devices, especially in Asia.

Financially it is still hard to derive solid overall return statistics from foreign investments into Asia, as investments are significant, operations are still in the early stages of development mode and only recently have started to pay-off. As we're still riding on an "investment high" in China and India few contrarians exist who bring realism to the present Asia hype, which is an issue which also permeates in the mobile and wireless space. Critical in this regard is that few Asian mobile and wireless `brands` have appeared which carry brand-equity beyond national borders. Except for hardware manufacturers and established media leaders like Sony, Nintendo and Huawei, few Asian mobile and wireless ventures have made it to the international landscape. Sony's move to appoint Howard Stringer as a first-time ever foreign CEO to reinvent the ailing consumer electronics giant is a sign that Asian companies are progressing towards a more international playing field and hence better chances for global brand equity.

Domestic and also regional content is being promoted, mostly fuelled by local telecommunication giants who also act as a springboard to mature small and medium sized content providers through a portfolio investment strategy. Such strategies can prove attractive once these digital content providers can tap into a larger international audience given the portability of their content across multiple platforms.

As national markets within Asia are under development, cross-border growth is expected through strategic alliances and joint-ventures, rather than via acquisitions, although equity participations as a means to a partnership are seen in publicly listed corporations. An example of this is Singtel's recent acquisition of a sizeable block of shares in Thailand's TOT (Telecom Authority of Thailand). The underlying rationale however is not the immediate economic merit, which would be found in Western transactions, but a long-standing relationship of trust between the two companies.

Hence investments need to be first considered on a national level and cannot be assumed to enjoy the rapid and smooth cross-border regional roll-out as can be experienced in a more homogeneous territory like the US. Hence, investors need to identify buckets of value, which will deliver sufficient investment returns to justify a transaction whereby an international roll-out is an extra premium.

Hot topics

- The digital content and gaming sectors are being actively promoted by smaller Asian economies like the Philippines, Malaysia and Thailand, which also facilitate their new ventures´ international business expansion through well-funded dedicated government agencies;

- Next generation technologies like RFID, Zigbee and WiMAX are all being explored and introduced ahead of commercial roll-out in the USA and Europe;

- Continuing large infrastructure investments continue to add to the investor appeal in these challenger economies we believe that this will set the scene for a breakthrough of a selection of these wireless ventures on the international radar;

- Meshed networks, given the different state of development of nations within Asia Pacific the meshed network technology provides for a good alternative to fixed telephony lines;

- Research and development is being discovered as the primary means to long-term sustainable growth, whereby relative inexperience within this field of smaller economies, compared to more mature economies like Japan, Korea and Singapore, spurs several improvements in policy, investment and tax climate and foreign partner promotion. Although a hot topic, the first fruits of this quest for R&D are expected to take some time, as the general infrastructure for R&D in the entire region is still considered to be at a very early stage.

What stage are companies at, in the wireless mobile space in Asia?

Given the high variety of maturity within the Asia Pacific region the general state of companies operating in the wireless mobile space needs to be reviewed on a country-by-country basis to avoid generalisations. In all, Asia has not only been dynamic in terms of growth; it has also been a leader in innovation with such products as NTT DoCoMo's i-mode in Japan and the application of SMS and widespread adoption of the market for ring-tones. With various emerging markets and a very high penetration rate of broadband services Asia is poised to become a development leader once R&D gets off age and becomes well-branded, driven by companies with an internationally scaleable business model.

All eyes will be on the developed countries in the region where 3G has now been officially launched and the progress that carriers will make in lifting ARPU through non-voice services. Critical for these markets are the availability of high quality actual domestic content to create local communities and interest groups, which ignite market adoption of data communication. The other end of the spectrum is set by economic conditions for data usage, which vary widely across the region and eventually determine sustainability of the model.

What do founders and investors need/look for?

As these reflections are general for foreign investors in Asian markets, some caution is needed for relevance for each specific regional territory.

Transparency of the local investment conditions, climate and business case can strongly influence the success of an investment in particular when illiquid investments in privately held companies are being considered. Investors have to rethink their strategies whether to pursue mature markets like Japan and Singapore, hypes like China and India or challengers with attractive value buckets like Thailand, Malaysia or New Zealand.

As with any investment case, the entrepreneurial team is the primary driver of success. Commitment, experience and network combined lead to the required chemistry that attracts business success, also in Asia.

Reliable local partner. The majority of mergers, acquisitions, new business formations or joint-ventures are being challenged by the existence of a reliable local partner which can smooth the market entry, access and expansion of the business in the target market. Although much attention is paid to financial and technical feasibility of a transaction, leadership, culture and common vision are the final determinants of any foreign market strategy.

                    Ajit Jaokar and Tony Fish

Formation of Asian technology brands

To date the global brands are owned by commercially more mature nations in the West. In the near term, a first set of truly Asian technology brands is expected. With Lenovo buying IBM's PC division and Ningbo Bird becoming a leading Chinese handset, brand centres for innovation are being created. Investors need to consider whether ventures hold the potential for a regional or global brand. Haier's failed, but daring bid to acquire Maytag indicates a more outward-looking strategy for Asian companies which eventually will prevail.

A regional cost advantage can be valuable, but this time it needs to be placed in a global context, whereby truly international enterprises start to emerge. With the recent advent of fast-track public listings on the Alternative Investment Market (AIM) in London, with increased popularity among Asian corporations, it is possible to foresee Asian ventures being transformed into Western corporations with a low-cost Asian manufacturing base and efficient access to capital.

Preparing the plan

Preparing and delivering the plan requires commitment! How committed are you and do you understand that the commitment is more than time, that is your life, your happiness, your family, your wealth and potentially your health. There's a real cost in the preparation of the plan and even more to actually realise it. Do you realise and accept that you do not have all of the skills necessary to grow the business to a billion dollars turnover? Are you critical enough about your own skills and do you know where to get those skills that you do not have and do you know how to manage those missing skills?

The quickest route to fail is to try and do it all yourself. Those that succeed depend on a wide network of experienced people, they heed warnings and they understand that it is not a black and white process, but they do lead. As such, these leaders are honest with themselves; this integrity comes through in the plan. While commitment is critical, honesty is the key.

If you are starting out and have never funded a projected before, you need the immediate help of two of the following three people, your father figure, your mother-in-law or your partner. The alternative is a very long questionnaire, but the difficulty with this route is that you would never believe the end result! Honestly!

The first hurdle is simple. If you are able to articulate your concept and business to two of the above people, they understand it and are positive about it; you pass the first hurdle. As in the 110m hurdle race, each hurdle is the same height at the same distance; it depends on your training and stamina to be able to jump them and cross the finishing line. What is interesting in funding terms is that it's not about winning it's about going the distance.

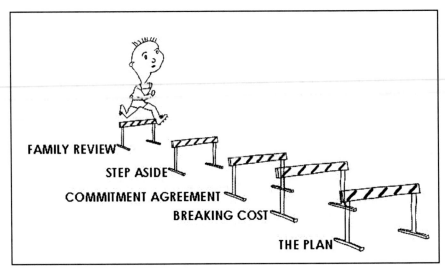

Figure 55: *Hurdles*

The next hurdle is itself a review that you cannot participate in or argue against. You should not ask this question at the same time as the first hurdle. The question goes to the same two people you've presented your plan to. The question is, "when do they believe you'll need to step aside as the senior person within your business to let somebody else grow and manage it?" You are not allowed to defend yourself. If you accept their view and you understand why they have that view, you're over the second hurdle. If you deny what has been told to you or you ignore it or you tell yourself that is not true you're not ready to continue yet.

Ajit Jaokar and Tony Fish

The third hurdle is to ask the same two people the next question, again, not on the same day and not in the same context. This time you want to know if they're going to stand by you through thick and thin and very, very thick and very, very thin. Are they willing to continually challenge you at all stages but always be there to support you? Are you willing to accept continual challenges from those closest to you, from those whom you need support motivation and encouragement? Once you have written an agreement (three lines only) that you sign with your partner, you're onto the next hurdle.

The fourth hurdle is one you need to do on your own. You need to determine at which point you're willing to quit. At what point will the cost be too high? These are the financial cost, the cost of the relationships around you and your health. You need to detail what you would sacrifice to make the business work. Having written this up, you need to share it with someone, as this will hold you accountable. If you cannot be held accountable or could not write a limit to your own personal commitment than you are not ready to proceed.

The fifth hurdle is the business plan itself. There are many good books and resources available on how to write a great business plan. There is no absolute to the plan itself, it is an articulation of what should happen if things go well. Within this context, there are five key elements that the majority of investors will look for. These are:- the story, the team, the numbers, the vision and the control of risk.

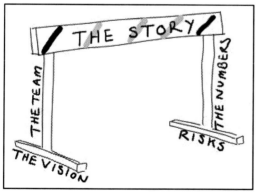

Figure 56: *The Plan*

a) The story: it should be short, concise and tangible. A prime consideration should be that the potential investor should at least have a basic understanding of the market that you're approaching. You should not need to spend a great deal of time describing the market, its size or its value, if you have to educate your investor, as a rule of thumb, you will never get there. The reason being that by definition the business proposition will be at a leading edge, these are not stationary business plans, and re-educating the investor as something changes will drain your resources. If there is a need to justify market changes and competitive elements, do it in a separate document or consider the market may not exist or it may not be big enough to attract funding.

b) The team: In reality, this is what the investor is backing. Time must be spent articulating the strengths of each of the individuals, how the skills complement each other, and how you work together, even a written agreement on how to deal with conflict. An important part of this is knowing when new skills will be needed, who will step aside and how succession planning is already thought through. There is a second part of the team who won't be listed in the business plan. These are the mentors and advisers to the management team. It is a good practice to have these as references. Additionally to the mentors and advisers, there are other family members and other wider network of supporters. Feedback on the proposition from this network provides supportive evidence that the management team is taking advice from the widest possible network, and not using a sample size of one, for all of their forecasts.

c) The numbers: To date we have not seen or been involved in a company who have fulfilled the numbers as presented in the business plan. They have been wildly better or worse! The numbers demonstrate the potential return, given an investment and forecast for income and costs. The most important part numbers play is to be an articulation of the potential size of the business and if the opportunity is there for a very substantial business, as this is what investors will be looking for. A key aspect of the numbers is the absolute but this only makes sense if the ratios such as income per sales person or overhead costs per employee make sense. These ratio tests are a good sanity check of an overall plan and there is a lot of good material online and books on business plans that would help you understand the detail. It will also provide the basis of reporting and management.

d) The vision: follows the numbers, as it is likely that the initial income sources will be different from the income sources that will make a significant return. The vision is how the company will navigate its way through the foreseeable future and how it sees the future and its position within the evolving markets. This part of the plan needs very careful thought along with the management team and numbers, as this is what the investor is buying into.

e) Controlling the risks: is the final important section within the plan. This part demonstrates that the management team understands and can articulate the risks and how those risks can be contained and controlled. This section would include operation and market channel risks and how you plan to deliver the plan.

The written business plan itself is not that important, however, it is critical that the key facts and underlying assumptions are well written and the words coherently describe a story about how the management propose to eliminate and control risk. Being bound to the precision of the numbers presented is dangerous, as is a management team (or investor) that will not deviate from the plan even though the market has changed.

Funding

The attitudes of investors towards a particular industry or technology vary as much as the attitude towards ties, length of skirts, music taste and the colour of your jacket. Essentially your plan will be read if it is fashionable.

The funding process is extremely individual; again, there are many good books and online materials that will provide specific details about the funding processes and what to expect. There are six key aspects which rise above all of the different methods and approaches. These six key aspects are:- types and sources of funding, presentation, diligence, expectation, advisers and commitment.

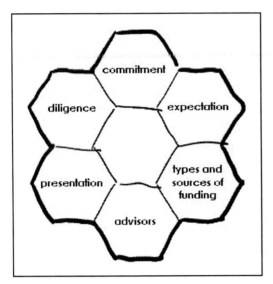

Figure 57: *Key aspects of funding*

a)   Types and sources of funding

The generic comment is that there are two types of funding, knowledgeable and dumb. Knowledgeable funding is money that comes from sources that understand and have a deep level of interest in your specific market or technology. Dumb money comes from investors who have a generic remit to invest into all sectors, services and technologies. Knowledgeable and dumb funding is available at every stage of the funding cycle. There are at least four stages in the funding cycle. These are your own with friends and family, seed investment, institutional investments, and public market investment. The type of funding that could be available to the venture is predicated with its current stage and growth potential.

It is worth noting that it is harder in many instances to raise small amounts of money (less than $1m) than it is to raise significant investment rounds (greater than $50m). There's an abundance of good material online that will help you find sources of investment, including grants from local government.

However, the key fact is that you must do your homework. You need to know who the investor has previously funded, what their specific levels of interest are and how much is left to invest from the fund. You need to know about timetables in terms of how long it takes them to make an investment decision and their sign off process. You can get much of this by phoning and speaking to senior management of companies they have already funded, always an interesting call.

As a principal, always start with knowledgeable funding sources.

b)   Presentation

A professional presentation is essential. Professional not only means well structured with a clear story, but also extremely well rehearsed. Before your presentation is given to any investor, you should have rehearsed the presentation at least twenty times. If you're bored by the twentieth time, you will be suicidal by the time you've done it 100 times, which is the level of commitment that is needed to get funding.

The reliability of your presentation is essential. What this means is that when investors speak to each other (as it is an incestuous market) they will speak about the same facts and figures, irrespective of when the presentation was given or to whom.   Each time you give the presentation you must use exactly the same words and not deviate off into something that is more interesting. This may sound rather harsh, but you only have one chance and it is no use trying to perfect the story in front of investors.

You must have the same enthusiasm for the business on the first presentation as the hundredth; this is a real test of team character. Your presentation should not change from the first presentation to the hundredth. If change is required to the presentation, you may consider starting all over again.

If you're presenting as a team, let every member of the team speak. Let no one person speak or appear more dominant than the others. The investors want to back a team not an individual. You must rehearse as a team and know where each member speaks, and even who is going to answer which type of question. Look like a team, behave like a team and win as a team.

---

c)   Diligence

One or more investors want to know more detail about the business, the management and the risk. Diligence aims to achieve two goals, the first is to determine if what has been presented is factually correct and the second to find areas of the management team skills where there are holes and how they are going to be addressed. Therefore, the underlying principle of diligence is honesty.

Diligence will cover at least technology, people, legal, financial and regulation or market controlled issues. On all aspects where you have control; open, frank and honest responses will help bring a deal together. An honest response is to admit that you don't know, but you know what to do about it, you provide a plan, you deliver it and it makes sense, this builds confidence. The investor does not expect you to know or do everything, but does want to know that what you claim to do, you can do.

If you have told any untruths, or overtly exaggerated, or misrepresented any details, contracts, contacts or income, it will be uncovered during good diligence. Consider this. If you get away with misrepresentation, it is likely to affect later decisions, or the funding company will not be as thorough as you would wish for, which could also cause you many problems later on.

d)   Expectations

There are two sides to expectations, what you expect in terms of timescale, valuation and investment, and what your funding partner expects in the term sheet and as a return. The term sheet will predominately be driven by the necessity of the cash requirement into the business. The more urgent money is needed the worst terms can be expected. The best funding therefore will be achieved when you don't need more.

The term sheet is a book in itself, but a key to getting the right term sheet is by setting your expectations out early.   It is worth noting that the investors need good investments as much as good investments need the cash from investors. It is symbiotic.   Investors are personally rewarded based on the performance of their investments. In this aspect, they have a conflict of interest. They need a significant proportion of the business to allow them aspects of control. However too much control, and they can disenfranchise the management team/ too little and they may not receive sufficient upside to justify an investment. (Dumb investors will act like parasites, leeching the host dry and moving to the next one.) Their consideration must also be that when further funding is needed their position may be weakened, or that the founders and other management team may have such a low potential upside they may not commit 100%.

It is the teams' responsibility to start open and frank discussion about these topics as will allow each side to manage expectation; this should be done earlier rather than later.

e)   Advisers

Advisers fall into the matrix below

	Knowledgeable	**Dumb**
Inflicted	Poor	Avoid
Chosen	Best	Care

Advisers can be inflicted on you by your investors or chosen by you and they can be knowledgeable or dumb. The worst scenario being inflicted and dumb. This is where your investor forces an overhead on the board. Conversely, if you're able to select a high value individual from a selection that the fund would be happy with, you can end up with a strengthened management team.

Ensure that you are able to select your advisers and how they are remunerated, what their deliverables are and how they report to funding source. If you do not manage this expectation, the advisor can work against you, instead of for you.

f)   Commitment

The commitment required has already been discussed earlier in this chapter. The commitment that is important at this stage is not a personal one to make the business happen, but rather the commitment to make sure that your time is spent equally and fairly between managing the team, delivery of the product and income, and the time needed to raise the money. Great care has to be taken to ensure that the team is still there and focused when the money arrives.

It is worth reflecting that personal commitment has two sides. There must be a commitment to make it happen, but the management must know when it is time to quit. The management will only know when it's time to quit if they have advisers that they trust and have made themselves accountable to them. This is why having inflicted and dumb advisors will allow a company to operate beyond the bounds of reason.

Numbers

Numbers, which translate into profit and loss and balance sheet, are critical. An infinitely variable Excel model that copes with every possibility, scenario and goal-seeking outcome is not critical. A set of numbers which can be managed, sets expectations, and focuses the mind is critical. Detailed micromanagement numbers and sticking to the exact number in the business plan for aspects of cost is not critical.

The number should be understood in the context of the ratios and the flash points. The flash points are twofold, one for the management to ensure all the things are going in the right direction, and corrective action is working, and to allow the management to report at a manageable level of detail without being tasked with justifying every penny spent.

Numbers are not only about income and cost control, but also about structure and return to the founders.

Excel is a very powerful tool, however it can be used to prove just about any business can make or lose money (cash). It is quite a simple exercise to adjust the revenue and cost lines to make them positive, without considering what the implications are. The link to real life is often missing. A simple example is that of sales people. There is no generic rule of thumb for the sales person to generate income as it depends on the product, service, platform, sector

Ajit Jaokar and Tony Fish

and pricing, however, Excel will allow the model to double revenues and reduce the cost of sales in a few simple clicks. The implication being that the expectation on the sales team may be unrealistic. Therefore, against each cost line it is worth considering the ratio between the cost itself and either the revenue line or the number of people in the team line. The same is true for salaries, by ignoring inflation, and bonuses it is quite easy to improve margin, however you may not have a team to deliver the revenue. The implication is that too much time in front of Excel allows the modeller to often lose sight of reality and practicality, a good guide, should the model go over 1Mbyte, it is too complex.

Two additional comments need to be made about Excel and modelling, indeed these apply to document and presentation files equally. First version control is essential. Someone needs to determine and control versions and releases of all documents to anyone outside of the business. This may seem rather picky, but should there be a change to the story or the numbers, it is essential to know who has what version, especially when you need to call people up and discuss details. Second, there is nothing wrong with deleting the existing version and starting over. In many cases, it will be worth building the final model from scratch again. To do this it will be worth drawing a flow diagram of the business model and starting over. Documents are often dramatically improved if written from scratch, as they will ignore the original version and words (and often a lot of detail that is no longer relevant to the new proposed plan and model).

In terms of the structure of the model, it should be built from two aspects that produce consistent results. These being the demand side revenues and the supply side costs. The demand side needs to start from a known and reference source for a total market size and then reduced to the specific market sector being addressed. In a similar vein, the price/ tariff will start from known reference sources and be modified and aligned to justify the position of the business case. Given the size of the market and the price, revenue is calculated. The supply side direct costs should be built from ratios of the effort required to sell the service/ solution, the channel-to-market costs and the delivery costs including IT, partners and purchase of connectivity. Overheads should be modelled and based on experience and these will be inline with what investors expect and the management team know and understand from previous businesses, the management's mentors will also provide a sanity check on this aspect.

Some business books suggest that two models, one for management (lowest cost possible) and one for investors (built in 'fat') is worth while. Experience suggests that one honest model is a better approach. The model should reflect all the best efforts of the team for accuracy, as experience also indicates that no business plan has ever been followed exactly, they are either totally and overwhelmingly overshot or appalling missed. The honest plan will determine quickly which way is up.

As a practical guide, the financial model that is being used for sales and cost management should not change more than twice a year. Once a model is fixed, even though things change, it should still be reported to, until there is entire management and investor approval to change. Sensitivity management is a very powerful tool for understanding the key sensitivities within the proposed business plan and model. The model should be flexed by changing every revenue affecting line by +/- 10% and recording the change in break even, period to cash flow positive and peek cash requirements. The same exercise should be repeated for the cost lines. Those input lines that affect the outputs by more that 10% should be watched and reported on in detail. As part of the sensitivity analysis, other cash affecting input should be flexed as well including debtor and creditor days and interest rates.

Scenario management should also be used to help determine actions and provide input into decision. Scenario management will allow major change to the proposed business structure, such as changing the ratios of sales revenues per sales person, or changing from a direct sales channel to an in-direct channel, types of finance available and structure. It is often useful to prepare a separate model for scenario management which will link to the final model. However, overall, focus less on making the Excel model sing and more on sales and bringing income into the business.

New shareholders

Managing your new shareholders requires probably more time than managing your senior team. However, the most important thing is what you do on the first day that the investment arrives. It is strongly recommended that you do the following:

- Invest all of the cash into a high interest bearing account. As the investment cash that has arrived will not all be spent on the first day, you will be able to put cash on deposit. It may only grow at a low rate, but it does grow.

- Negotiate with all outstanding creditors. Cash should not simply be used to pay off creditors. A call should be made to each one offering a payment plan and maybe a reduction in the value. Once your plan is agreed with a creditor, don't fail to pay, as this would be very unprofessional. While this is hard, it will send a clear message to all staff that cash is never to be wasted.

- Inform the team now that only 60% of what actually arrived, has arrived. While this may seem a little dishonest and it may prevent excessive spending; it does provide a trigger for raising new capital. The trigger being, when the 60% cash has been spent, the provision of 40% allows time to conclude new funding.

- Try and arrange an overdraft facility at the bank. As the business will have available cash, present to the bank manager a plan that shows an overdraft requirement in a 6 to 9 month window. While there will be no ambition to use this facility, having it available may prevent accepting poor term sheets at the next round due to the lack of cash to continue operations.

Now that there is cash in the bank don't behave differently from the period when there was no cash available; do not start paying creditors off, except for the preferential creditors such as duties and tax.

The hardest part is that the team is allowed only a few weeks to get back to managing and building a business (and maybe a holiday) and then immediately start planning the next funding round.

It is very important that good communication practices start immediately with your new shareholders and the board meetings and other communication meetings are put into everyone's diary for at least the next twelve months.

Board meetings and board notes should be professionally run and taken respectively. It is worth engaging with someone who has run board meetings to establish good working practices and structures. This could possibly be an independent non-exec.

Web 2.0 Business models

A discussion about Web 2.0 Business models

Introduction

With all its emphasis on community, social aspects etc; one may wonder if Web 2.0 has a commercial aspect? In fact, the question of 'Where is the money in Web 2.0'? ..is quite common. On the other hand, venture capital firms worldwide are investing heavily in Web 2.0. In this chapter, we shall discuss the revenue models behind Web 2.0 and the reason for its interest to the investment community. Note that this chapter covers Web 2.0 in general and not Mobile Web 2.0.

The next time anyone asks you the question: 'Is there any money in Web 2.0', remind them (and yourselves) that the greatest money making idea on the Web is based on Web 2.0 .. We all use it. We are all addicted to it. We can't do without it! And it's not even patented as far as we know. And that idea is.. PageRank. PageRank[195] is perhaps the best example of building and monetizing a hard to create data source. Let us consider why PageRank is 'Web 2.0'.

---

195 http://www.google.com/technology/

PageRank as Web 2.0

We discussed previously that you can view Web 2.0 as the Intelligent Web. In fact, if you switch the order of the seven principles and consider that the second principle (Harnessing Collective Intelligence) is the core principle and that the other principles feed into it, then Web 2.0 becomes a lot clearer. Now, consider PageRank in the context of harnessing collective intelligence and the seven principles of Web 2.0.

a) PageRank uses the Web from first principle (Web as a Platform).

b) It uses data as its core element (Data is the Next Intel Inside). Data in this case, is the millions of links that it tracks.

c) There is no software release cycle. It's all updated continuously and dynamically.

But most importantly, PageRank 'harnesses' intelligence based on user input driven by the Long tail (millions of small contributions from users as opposed to a few significant contributions from 'important' users) and creates a 'hard to acquire' body of knowledge plus a means of monetising it (i.e. the body of knowledge thus created by PageRank is valued by the users). So, to us, PageRank is an excellent example of a Web 2.0 service.

So, why is there so much doubt about monetisation of Web 2.0? Firstly, there is a lot of hype. There is not much we can do about that except to know that Web 2.0 is more mature than Web 1.0 (and we are all wiser as well!).

We discussed PageRank as an example of how to create a unique digital asset. In revenue terms, perhaps it is easier to tell 'what not to do' to reinforce the concepts of Web 2.0 revenue models. To explain these ideas in terms of familiar concepts, we have used the example of a Digital community. Most of us have, at some point, joined such a community. By 'community' here we mean, members have a profile, they can chat, they can rate profiles/content etc.

Communities! = Web 2.0

So, is such a community 'Web 2.0'? Our view is: No. Here's why.

No tail:

Web 2.0 is all about monetising the Long tail. Often, community oriented sites do not monetise the 'Long tail': To recap, Long tail needs lots of small contributions from many users (often contributed implicitly). To do so, the site must:

1) Firstly, attract the Long tail (many members).

2) Secondly, capture content from the Long tail in a form that can be aggregated.

A membership of thousands can't lead to a valuation of millions! You need the numbers make money from Web 2.0 (i.e. the Long tail).

No Harnessing of Collective Intelligence

Merely having the numbers is not enough. A site may have the numbers but may not be harnessing collective intelligence. Comments on blogs/articles don't count unless they can be aggregated into something valuable to the customer. After all, let's not forget that an 'Amazon review' is a comment. But there is a huge difference between 'a comment on a blog entry' and a 'review on Amazon'. For starters, reviews on Amazon can be aggregated into ratings (Nine good reviews out of 10). Further, reviews on Amazon can be 'discovered' easily (for example against a specific book). Finally, and most importantly, the customer values those reviews; either in their raw form (text) or in their aggregated form for example: 'Four stars'. The customer may not always value simple comments on a blog/site.

Going for the 'head' at the expense of the 'tail', but still headless ..

When sites do not monetise the 'Long tail' they attempt to monetise the 'head' i.e. the 'few' rather than the 'many'. This is often at the expense of the Long tail and is an attempt to apply an offline model to an online world. On first

Ajit Jaokar and Tony Fish

impressions, focusing on a few paying customers sounds like a good idea. After all, the few who are the 'target customers' are 'high net worth'. They have money. So, surely they should pay a 'premium'. The problem here is two fold:

- Firstly, the few (head) do not necessarily want to go 'online'; not at prices to support the site's business anyway and

- Secondly, from the site's perspective, focusing on the head is often at the expense of the Long tail.

Focusing on the few (head) is an offline business model driven by scarcity. We believe that it's a hangover from the dot com days. In the later half of the dot com years, people insisted that sites were B2B, whereas B2C was considered a pariah. In the same way, people have a (false) sense of security when they think that they have identified a group with 'money'. Whether that group is interested in paying you any or not is altogether another matter!

So, a model of targeting the 'head' does not quite work and it is actually contrary to Web 2.0. Not only is the 'head' not interested in paying and may sometimes not see the value, (think diamond merchants) but more importantly; if you orient the site towards the head; **you cut off the tail**. For instance: if you wanted to work with a few, large customers; you would create agreements and contracts and have a high joining fee for the site. All this will alienate the vast bulk of members (the Long tail).

Closed systems: restrictive memberships: For the same reasons as above (it targets the head and not the tail).

No mechanism for quantifiable implicit contributions: See section on 'No harnessing of collective intelligence'.

No aggregation of user feedback: See 'No harnessing of collective intelligence'.

Proprietary technology: Restricts the membership growth and restricts interaction between members.

A hidden agenda (slapping a fee): This is a valid Web 1.0 model, but not a valid Web 2.0 model. The script goes something like this: Start a 'free' site, invite many people, Create buzz. Build up the numbers. Insist that you are 'free' but plan on slapping a membership fee for premium services. This is a valid model but it is not Web 2.0 in itself if there is no harnessing of collective intelligence, catering for the Long tail members, etc.

Too narrow: Targeting a small subset of the population (a specific country, group etc) rather than ideally, the whole world!

Social contributions without a goal: User generated content/social contributions have huge potential; but without a goal, they are just 'a lot of noise'. For instance, everyone understands what 'Linux' development is. Not so common perhaps is antbase.org ('A site that does for ants what bird watching has done for birds), which is also excellent. In both cases, there is a clear goal. Without that goal, you have noise. Lots of contribution, but no real direction.

So, what works?

Here are some ideas that are valid Web 2.0 revenue models. We have referred to some of the excellent work done by Dion Hinchcliffe in this space and specifically his blog 'Creating real business value with Web 2.0' 196

The resurgent adverting model driven by rich media and broadband: Once dismissed, the ad sponsored model is back with a bang! The software industry is about a $40 billion a year industry. Advertising is worth about ten times as much and its all moving online (hence witness newspapers in financial quagmire).

To re-emphasise, the monetisation of the Long tail is critical in a Web 2.0 business model.

Gaining control of a hard to create data source and then being able to successfully monetise it: For example: www.craigslist.com or PageRank.

---

196 http://blogs.zdnet.com/Hinchcliffe/index.php?p=14

Ajit Jaokar and Tony Fish

Web services including Mashups: Amazon.com is the best example. Currently estimated to be earning Amazon $211 million a year. The Amazon web services model 'decentralises' the store by converting the store itself into an open, easy to use platform.

Serving the large number of small customers through using the Web: Examples include SugarCRM [197]and salesforce.com. Note the key difference in this case is; the business model is oriented to serving the 'Long tail' customers. Thus Siebel [198] may also claim to serve smaller customers, BUT Siebel's core business model is designed to serve the large customers.

Give something valuable for free: This idea comes from Nat Torkington from the O'Reilly web site (seven principles of Web 2.0)[199]. *Another way to look at it is that the successful companies all give up something expensive but considered critical to get something valuable for free that was once expensive. For example, Wikipedia gives up central editorial control in return for speed and breadth. Napster gave up on the idea of "the catalog" (all the songs the vendor was selling) and got breadth. Amazon gave up on the idea of having a physical storefront but got to serve the entire world. Google gave up on the big customers (initially) and got the 80% whose needs weren't being met.*

Finally, there is the question: Should Web 2.0 have a revenue model as the VCs see it? (which is not to say that Web 2.0 does not have a revenue model). The Web is a great leveller. This means, many more people will become financially richer but it will be difficult to be very rich. With Web 1.0, inspite of the boom and bust, many countless people actually made money and created a viable business.

Indeed, although very common, PageRank is so unique that, it is almost impossible to emulate. The success of PageRank indicates that Web 2.0 in general, is oriented to creating big winners. Hence, VCs will always be attracted to it because Web 2.0 companies have a clear barrier to entry if they become dominant players in their sector.

---

197 www.sugarcrm.com
198 http://www.oracle.com/siebel/index.html
199 http://www.oreillynet.com/pub/a/oreilly/tim/news/2005/09/30/what-is-web-20.html?page=4

Salt, Pepper and Social Business networking

This chapter discusses the 'free' model in terms of Social business networks.
It has some implications for Web 2.0 and Mobile Web 2.0 because Web 2.0
and Social Networking have some common elements.

Introduction

This article discusses our views on business models for social networks. We
believe that Social networking should be like 'Salt and Pepper', a condiment:
free or very cheap (the salt and pepper reference was made originally to WiFi
as we see below). The real 'value add' is the value created by people using
the network (i.e. nodes) or ancillary services that can be sold on the network;
rather than the network itself. The network's financial viability is then based
on one or more secondary revenue models. We are seeing a similar effect with
Web 2.0.

Social networking is a relatively new phenomenon. We are using this term here
to also describe the sites included under 'Social business networking' (which
also use the principles of Social networking). There is no shortage of theories
about social networking[200], but oddly enough, none of them talk of money/
revenue models for social networking sites. On first glance, that sounds very
unusual: *until you realise that social networking sites are a microscopic
version (i.e. model) of the Internet itself and are simply mirroring the
Internet.* This is a strong theme in this chapter i.e. Social networking sites
should conform to the ethos of the Web; else it is akin to fighting gravity and
it won't fly!

WiFi as a condiment (salt and pepper)

Back in September 2003, when we were all still in the nuclear winter of the dot
bomb, there was a short, insightful article in Wired magazine called: 'Would
you like WiFi with that' by Paul Boutin[201].

---

200 http://en.wikipedia.org/wiki/Network
201 http://www.wired.com/wired/archive/11.09/start.html?pg=4

Ajit Jaokar and Tony Fish

It asked the question:

> *If wireless Internet access is such a hot technology, why is it such a dud business? Wi-Fi hardware, which uses radio signals instead of cables to connect computers to the Net, is already in more than 10 million laptops. But try to make a buck selling connectivity.*

Sounds familiar?

But then, it goes on to provide a solution:

> *If you want to see the right way to serve wireless access, find a Schlotzsky's Deli.*202

> *The Austin, Texas-based sandwich chain figured out the secret of making money from Wi-Fi: Give it away. Schlotzsky's lets anyone sign up and use its network free, even if they don't come in for a sandwich. The chain advises its 600 franchise owners to beam Wi-Fi signals through the walls into nearby hotels, parks, and college dorms. Such complimentary access points are popping up everywhere, from Buck's, a roadside restaurant in Woodside, California, to the Portland Harbor Hotel on the Maine coast. And why not? Giving away wireless broadband saves on billing costs, attracts customers, and creates an instant competitive advantage. Buck's owner Jamis MacNiven, who serves buttermilk pancakes to some of Silicon Valley's top venture capitalists, has the perfect rap on the topic: "Charging for online usage would be like charging for salt and pepper."*

Read that last sentence again: **"Charging for online usage would be like charging for salt and pepper."**

And the article ends with: **Wi-Fi isn't a luxury or even a commodity. It's a condiment.**

Social networking as a condiment and the Ethos of the Web

Substitute WiFi with Social networking in the above discussion and you see why so many of the social networking sites have a business model problem. Many of these sites are contrary to the ethos of the Internet. That's not to say

---

202 http://www.schlotzskys.com/

that they don't have a business model, or even that they may not be making money. It's just to say that they are retrofitting an offline business model into an online situation and that has limitations.

And what is the ethos of the Internet? Its something we have been advocating to Mobile Network Operators for years now: You can summarise it by the phrase: *'Dumb pipes and Smart nodes'.* The network (Internet) itself is 'dumb'. *'Its only job is to 'connect people'.* The value is provided by the nodes (the people / systems that are at the ends of the pipe).

Jonathan Schwartz summarises these ideas in the 'Power of the end nodes[203]' (AKA: 'the network is the computer'). We believe that the same phenomenon applies to social networks on the Internet(and by extension, to Web 2.0). The moment you introduce tiered membership, complex pricing models and so on, you hamper connectivity. The effective size of the network decreases because all members can't do all things. 'Dumb pipes and Smart nodes' is how the Internet has always worked. Hence, witness the uproar of the net neutrality folk [204]and comments from Sir Tim Berners Lee[205]. And rightly so! The same applies to WiFi and the same principle is also applying to social networking(and Web 2.0). *This is the social equivalent of the digital concept: 'All packets are created equal'.*

Observations and recommendations:

a) Subscription model or a tiered pricing model is a valid model. But, it is more a remnant of offline business models than online business models. The point simply is this: you can't build a 'web business' by restricting people. We can build 'a business on the Web' but that business model is limited because it is retrofitted from the offline world into an online environment; where it does not have a natural home. The same logic extends to other web models like 'Social production'[206]. We can't selectively apply one aspect of the Web such as 'social production' and

---

203 http://www.alwayson-network.com/comments.php?id=13472_0_11_0_C
204 http://www.savetheinternet.com/
205 http://news.bbc.co.uk/1/hi/technology/5009250.stm
206 http://www.amazon.com/gp/product/0300110561/104-8061864-5624728?v=glance&n=283155

Ajit Jaokar and Tony Fish

ignore others i.e. 'its inclusiveness'. Such a hybrid model is not valued by the investment community (and rightly so: because it won't scale)

b) Could we perhaps argue that we have a smaller number of members but each of our members is somehow more valuable than those in other networks? Interesting, but again not a model that mirrors the ethos of the Web. It's like saying 'packets in my network are more valuable than the packets in your network'. The online equivalent of 'Mine is bigger than yours'.

c) As I have mentioned before, the advertising model works assuming you have a large, flat/ unrestricted network. Advertising is more than 'Google ads'. Its more driven by richer media and broadband in addition to text based ads. That's why VCs favour the MySpace model. Note that this model (for instance MySpace) conforms to the basic Internet ethos.

d) A network should be like Nokia[207] - connecting people! *If it does more (i.e. tries to act smart), then it starts to disconnect people!*

e) Nodes may not be people, they could be an aggregation of people i.e. clubs. i.e. a group of people creating a 'network within a network'. The club could be built around a specific purpose. It is not the network; rather it is an endpoint in the network with a synergistic relationship (and potentially a revenue share relationship) with the core network.

f) One service to watch is Marc Canter's 'PeopleAggregator'[208], and by extension the whole idea of open social networks because they follow the principles of the Internet. In the case of PeopleAggregator, each node (which is a separate network in itself) could act as a node within a much larger network.

---

207 http://www.nokia.com/
208 http://www.peopleaggregator.com/

As per the people aggregator site:

> *At the centre of all great Live Web software will be one's digital identity*
> *Each end-user will control who has access to their personal data, who can*
> *use it and how. Each end-user needs to be able to move their data to other*
> *systems – whether it be their personal profile record, list of friends, groups,*
> *their photos, MP3s, links, bookmarks or events.*

Oddly enough, open social networks have a business model, exactly for the reasons we mention above i.e. conformity with the ethos of the Web. For instance, in the case of PeopleAggregator,

a) For PeopleAggregator itself, it's a 'software sales' model.

b) For each participating network, it's the possibility of attracting more members and by extension, each participating node is likely to be specialised.

c) For the users, it is freedom from one specific closed network and the chance to meet more people globally.

Conclusions:

If you have followed the arguments so far, it leads to the conclusion that the **'Internet itself is 'salt and pepper''!** And that's exactly spot on! The Internet connects people. Complexity and offline models disconnect people!

We are not necessarily advocating a free social business network, but definitely a flat, inclusive, open, 'web like' social business network whose primary job is to connect. Such a network follows the basic ethos of the Web and is likely to be based on a secondary business model. This business model could be advertising but need not be. Whatever that model, it should not cripple the primary purpose of the network

Ajit Jaokar and Tony Fish

What's on the mind of the CEO

This section presents the key topics on the mind of the CEO across a wide selection of industry sectors. The sectors we have chosen are being affected, or are going to be affected, by the Web 2.0 and Mobile Web 2.0. This section covers key strategic issues and not specific issues by company since each company will have different concerns and motivations. Thus, while not specific, the topics and ideas represent the key themes that are driving certain industry groups. The topics were constructed following a significant number of informal and undocumented phone calls and interviews. Many companies are active in more than one sector (silo) since there are few global single sector companies. This adds to the complexity of the analysis.

Some CEOs view Mobile applications as purely an extension of the web and others as an access technology. Yet others as a stand alone industry, protected by airtime licenses and technology. The key benefits from Web 2.0 and Mobile Web 2.0 will be gained by those CEOs who take an opportunistic stance i.e. they see that Mobile applications and the web are symbiotic with distinct and unique values that need to learn and grow together and that the advertising model is broken. We depict this in the following tables.

Network operator	Vodafone, Verizon, DoCoMo, Deutsche Telecom
Strategically important changes occurring:	
Declining RPU	Operators of both fixed and Mobile Networks face increased competition on tariffs for domestic and international voice and data.
International Roaming	Major issue for operators as consumers and regulators have shown significant interest in the very high costs of calls to international destinations and calls when a handset is receiving and originating calls in a foreign country,
VoIP	Voice over IP. This provides an interesting entry point for competitive carriers to enter into domestic markets without the need for infrastructure and the ability to market cheap or free calls.
Fixed mobile substitution	FMS, this is the migration of the traditional fixed line subscriber to mobile only. The fixed subscriber stops using a home phone line and instead only has a mobile phone.
Convergence	Convergence itself could provide a host of opportunities, however, the technical costs and implications on existing legacy back office systems are still not quantified and hence are a concern for many operators.

Long term business issues:	
Applications (Value Added Services)	The major issue for operators is that they could become an access layer only unless they develop or become a high quality supplier of value added services. However, this is not their core yet and in comparison to revenue and profitability from voice and data services, VAS are not that attractive in the consumer sector.
Wholesale	Most operators offer a retail and a wholesale offering. Their wholesale offering allows retail competitors to compete with their own retail brands. This is a very difficult issue to manage and to enable separate decision making on key issues, as they will in most cases affect each other.
Payment methods	Payment for services is a prime long term issue. Existing billing and collection arrangement allow operators to enjoy a billing customer relationship. However, as users take up more value added services, these may not be billed via the operators billing platform, as the user will have a choice. This affects the user view of the operator (which could reinforce an access only perception) or that the operator is expensive as the user has many services billed for, via them. This in itself introduces issues about warranties, returns, call centres and customer care.
Strategic opportunities to balance major concerns:	
Mobile Web 2.0	**A network operator who picks and develops devices, access methods and applications will be able to maintain and indeed improve a premium rate service, as the service offering will be user focused. However, Mobile Web 2.0 will have little effect on the key decisions involving change to the fundamental infrastructure and build.**

Solutions developer	Microsoft, IBM, Accenture, Nokia, Cisco, HP, Matsushita,
Strategically important changes occurring:	
Mobility	Mobility in this case is the mobilisation of the work force or the solution set to mobilise a work force. This has effects on asset [property] management (less people in the office), IT systems and working practices. The upside can be that the field force can have the information at hand and the reduction in input time and re input errors.
Security	Security of data on mobile devices and the transmission of data over a wireless network causes a fair amount of anxiety within corporate IT solutions providers. Whilst important to get a good security solution and policy, it is difficult to police and manage.
Skills	There is a shortage of skilled people who are able to design, build, implement and manage integrated/ converged solutions. The additional skills/ knowledge for back office function integration is ever increasing, as is the issues from solutions being built on the infrastructure of different yet similar providers.
Design methodology	Implementations of new services and corporate applications with a mobile component are still bespoke. While there is some repeatable work, a majority of the design and implementation is unique, this has a dramatic effect on pricing and therefore prevents many corporates undertaking a full scale implementation.
Channel	Channels to Small and Medium enterprises are weak, however, these companies appear to be the early adopters of mobile enterprise solutions. The larger SI are finding it difficult to service this market both in terms of access and pricing. This is leading to the development of specialist new SI, who are developing skills in this new core area.

Ajit Jaokar and Tony Fish

Long term business issues:	
Partnerships	Due to the complexity of the solutions and the level of legacy solutions, there is a need for partnerships with both network providers and manufacturers to fulfil the requirements. However these partnerships need to be formed on a case by case commercial basis, rather than a long terms strategic one, this is different to current expectations.
Variance by country	The complexity introduced by cross boarder implementations and design variation are vast, especially when different equipment and network providers are needed in each country.
Scale of projects	There is a both an issue with scale and volume. There are very few large scale projects and far too many small scale projects. This combined with skills and pricing make it difficult for the larger SI to work on many mobility projects opening the market for small specialist companies.

Strategic opportunities to balance major concerns:	
Mobile Web 2.0	**Mobile Web 2.0 has a lot to offer in terms of applications and simplicity and will be an important for as designs mature.**

Brand owner	Unilever, Coke, Levies, Nike, Ford, Samsung, Adidas, BA, Cadbury Schweppes,
Strategically important changes occurring:	
Advertising model	The traditional methods and approaches are becoming obsolete and currently the replacements allow for KPI performance measurements, but may not be the solution. Ad companies are not reacting fast enough and thought-leading companies will take back advertising functions.
Decline in TV	TV has been a stalwart of reaching the public for Brand Owners. With TV in decline and the nature of TV behaviour in flux, the messages on TV will change, that leaves a challenge for brand reinforcement, building and product promotion, sales conversion.
Growth of on-line	The challengers can re-produce equal products with no brand, viral marketing all online. These lower cost operators can target brand owners markets with devastating effect, but the opportunity is to catch up and indeed lead, however this produces channel conflicts and pricing difficulties.
Retail costs	Retail infrastructure is expensive and increasing questionable in terms of scale. However, the benefits of 'touch and feel', customer care and returns provide a good basis for some level of retail support.
Decline in loyalty	Loyalty has become temporary, indeed to the point of a single purchase. However, while continuous repurchase of one product as an immediate replacement may be in decline, repurchase over time is not. The understanding of customer ownership and loyalty needs to grow up, including who owns the customers preferences.
Price comparison	Once hard and difficult, now possible and will increase, becoming the norm, including at point of purchase, via mobile devices. This will change consumer behaviour and indeed pricing decisions/ willingness to negotiate price. Whilst a direct effect on margins and returns, it is difficult to increase pricing, due to low cost operators and declining brand loyalty.

Long term business issues:	
Differentiation	Consumers are more aware than ever that true differentiation is marginal, but usually more costly items are of a high quality with better service. Therefore, decision is to be lowest cost producer or niche high quality!
New product development/ IPR	New product development is expensive and patent/ IP is difficult to protect. The long-term issue is to be able to engage the consumer base in the development and indeed getting the consumer to actually do the development.
User generated reviews/ rating	Potentially the most interesting long-term change will be the direct feedback and access of reviews from consumers. In the case of a company with good products and high standards of after sales care, reviews will be the greatest recommendation. However, there will be a gap between the investment to deliver this and the time to create the viral support, which may force some brands down the lower cost route to maintain returns.
Strategic opportunities to balance major concerns:	
Mobile Web 2.0	**Mobile Web 2.0 will have a significant effect on Brand owners, as there will be changes to the interactions with customers, how and where interaction occurs and the availability of competitive services and information.**

Ad agency	WPP, Omnicom
Strategically important changes occurring:	
Changing business models	The traditional business models that are well understood are changing. Their revenue streams look to be decreasing and therefore there may be an unwillingness to follow that path. However, there is a need for increased creative talent.
Understanding Internet models	Agencies have got to move from following at a distance to being at the cutting edge if they want to be of most value. This means adopting and trailing far more new ideas than traditionally. This, however, will lead to many failures and some creative souls will find this level of failure hard and the increased level of support to test out new economic models will bring a burden on costs and margins.
Understanding mobile opportunity	Mobile and Mobile Web 2.0 provides a change and a challenge. Agencies can no longer wonder about how mobile fits into the mix, it is becoming the driver of the mix. The five Ps becomes old world.
Creative pressures	There is a need for creative ideas to be down scaled from HD TV and Video to 160 characters and IM. Simplicity will be the new creative, however this will not feed the creative spirit and therefore causes a management of skills issue.

Long term business issues:	
Even playing field	Significant value could be obtained through global relationships with content owners and TV channels for the timing and distribution of adverting. These relationships are being eroded by the Internet open and even playing field and the change in TV viewing habits. Significant Internet players themselves have to maintain a high degree of innovation to say still.
Excellent online tools	The development of tools to manage, monitor and report on key performance, immediately, are certain to differentiate agencies in the future. Indeed, client tools that allow the client to modify, change, update and design will emerge challenging all value-added services with the exception of creative.
Measurability	There is a growing expectation that everything can be measured, which for many aspects of the new marketing mix will be true. However, the leading agencies will develop services that will be outside direct measurement by KPIs, allowing such agencies to maintain a premium role. These will be tools, relationships, innovation in cross media delivery and technology innovation.
Strategic opportunities to balance major concerns:	
Mobile Web 2.0	**Mobile Web 2.0 provides the opportunity for Ad Agencies to change their business model and lead into a new generation of ideals and customer interactions.**

Internet Service Providers	Orange, AOL, Easynet
Strategically important changes occurring:	
Access only business model **NGN roll out**	Users will buy broadband as access only, as the bundled portal services have a declining value. Bundled broadband and voice (mobile/ fixed) is only an access model and will continue to be under intense regulatory review and indeed price pressure.
Alt net wireless offers	Next Generation Network roll out allows new players to enter into the broadband retail market without the need for capital and infrastructure. This will change the dynamics of ISP and the service offerings, again. NGN will enable the rapid expansion of international roll out for all players.
Decline in "banners"	Alternative access solutions from wireless technologies enable a different offering as a bundled mobile offering on one access technology becomes possible. Single pure play could have a cost advantage. Whilst the banner ad has enabled a new revenue stream, it is likely to pass as quickly as it arrived.

Long term business issues:	
No differential advantage	Email, storage, IM and portal services have no advantage. Indeed users will want a trusted safehouse for their digital identity and the ISP will be an interim step prior to the users wanting that as a separate service. Customer service will be important, but do users value this as the cost will be high due to the complex nature of home IT solutions and the variation in equipment/ setup and user skills.
Unsustainable long term access model	Access only is a scale marginal play that is best suited to the lowest cost provider, whom is likely to be the NGN infrastructure provider, however wholesale regulations will arterially keep competition active and may prevent the best consumer deals and bundled offerings will continue to confuse the
Wireless technologies	consumer. Wireless in countries where there is no or little copper infrastructure is very persuasive. Wireless could allow ISP to expand and own their own infrastructure in emerging countries and enjoy an access model with sustainable margins.
Strategic opportunities to balance major concerns:	
Mobile Web 2.0	**Mobile Web 2.0 provides the opportunity for ISPs to change their business model and create a new user service level. Having lots of registered access users or email addresses will not be a long-term value play.**

Service company	SKY, ABN, Swiss Re, Sun Life,
Strategically important changes occurring:	
Brand awareness	The generation whom are buying services today have grown up with TV and advertising. The generation following us will be different. Reaching this new generation with products and services and creating loyalty will be a challenge. This new generation do not like technically limiting rules, and those that try to control in the same capacity we have, are likely to be sidelined. Addressing 'Openness', in its fullest concept, will be the challenge.
Channels to customers	The underlying move from traditional channels to online is ever present. This new channel opens new issues with marketing and the scope of reviews and feedback from customers on your product and service. Service companies who excel in customer care, and pay attention to user generated reviews will build relationships, but these will be different to the loyalty experienced since 1960 to 2006.
Bundles	Service companies increasingly look at bundling as a mechanism to add value. ISPs experience in bundling is that users eventually pick out one service that it of value and perceive others as commodity, and will buy at the lowest price. Therefore bundling is not a long-term option, but may maintain the momentum of change. However, in the absence of a truly informed market, pockets of opportunity will always exist.
Operational costs and efficiency	The drive will continue to push cost out and improve efficiency, however, service providers have to face innovation and R&D at some point, as cost reduction can not continue at an every increasing pace to maintain the expectation of the City. The issue being at what point does innovation become the lifeblood?

Ajit Jaokar and Tony Fish

Long term business issues:	
Crossing the chasm and tipping points	Great concepts/ books with important implications for service providers. Given that the Internet provides a very cheap channel to market and open source technologies allow for rapid and near free development, Service Providers have to focus on products and services that will be mass market. There is an increasing pressure to just try, rather than follow a well-structured process for product development and introduction. The long-term issue is balancing innovation and rigor. Too much of either will overload or slow down respectively the company.
Management	Management skills for the management of change and control of innovation are in short supply as are board skills needed to provide the direction.
Outsourcing	The question is not that outsourcing makes sense, or what should be outsourced. It is more a long-term question of what business are you in. Given that service quality will become increasing critical for success, indeed more important than the product or service. Is service creation the core business or servicing the creations?
Strategic opportunities to balance major concerns:	
Mobile Web 2.0	**Service companies need to understand Mobile Web 2.0 as every part of their business could be affected. Whilst not the instigators of the technologies or underlying services, they will be able to provide the mash-up services to the real customer for payment.**

"Search company"	Google, Yahoo, Yell, White Pages, iTown, InfoUSA, DFN
Strategically important changes occurring:	
What is the business model	The value of searching, matching and finding are still largely unknown, as indeed is how users will be willing to pay. It is evident that the existing models are open to abuse and that companies can buy their way into top positions, which is opposing the unwritten Internet Rules. This in itself will force change as disruptive developers will develop and distribute tools to ensure that those who create rules and use them for commercial benefit will have only a limited time for exploitation. The business model is not stable.
Database of intentions	To understand the intention of the input requirement is driving technology and a need to a digital identity. However, the user will want to control their own digital identity and access to it. It needs to be explored how many consumers will allow a search provider to store personal preferences that the provider can exploit for commercial benefit.
Metadata	That embedded data will prove as valuable as knowing a user and their preferences. Creating metadata on voice and video will be critical for the next stages of growth.
Win or lose	Like the underlying code, win or lose in search will be binary. The Internet channel of near zero production and distribution cost means that there will be a continual change at the top of the league as sustainable advantage is difficult as a pure play model.

Long term business issues:	
Migrate to service company **Subsidiary for other business areas** **Speed to change**	Search, even directory services will come under increased pressure due to free competitive services. The opportunity is to convert search into a service, such as looking up a person, but then being given the choice of connection, rather than the one that your phone provider would offer. Search is a core technology and service and will become increasingly important. However, what are the key services that it should provide access for? The long-term issue will be preparing for the month when the whole model changes. It is unlikely to be a gradual change over time as has been historically the change in corporate life. It is likely that something extremely disruptive will force change over night, and therefore being prepared and indeed forcing such change will be critical to long-term success.
Strategic opportunities to balance major concerns:	
Mobile Web 2.0	**Everything is to play for with search and directory companies. Ownership of the database is good, adding value is essential, but extending this to deliver service will be the essential interruption of Mobile Web 2.0.**

Content Owners	EMI, Disney, Sony, Reed Elsevier, TF, Economist,
Strategically important changes occurring:	
DRM/ Rights by 'type'	Content owners have always enjoyed rightful ownership and exploitation of their materials. Over time, the right to exploit the content has become an increasingly exact legal art in terms of where, when, how and for how long. However, consumers are more confused than ever about what they are buying. It is likely that users will increasingly believe that they have rights through purchase, but cannot comprehend that they will be unable to use this content in whatever fashion they choose by DRM implementations.
Changing business models	DRM appears to control rights and present lots of opportunities for further exploitation. However, the disruptive generation, currently our youth, will not engage with companies who will be limiting their ability to exploit content they have purchased. This will force a rethink of the business models including the role of online retailing
Online business	Online introduces many opportunities especially in the domain of user generated content. This will be a source of innovation and feedback for content owners. Further, as editorial will be needed to make sense of the volumes of user generated content, it will allow those with such skills to exploit their own and user generated content to create a more powerful and compelling range of interesting services.
Retail issues	Retail has been a significant outlet for content and the changing relationship that consumers have with the high street and out of town facilities means that content owners will find it less attractive. Glossy images provided a mechanism for increasing sales, and whilst it will never disappear, it does not translate. The experience of browsing physically has not moved online as yet.

                                  Ajit Jaokar and Tony Fish

Long term business issues:	
Declining barrier of entry for new brands	As with anyone offering services, online has reduced the entry barriers. It will be possible for a content owner to only offer online movies, newspapers and magazines. These will emerge as technology improves, as will usability.
Device and device access	Devices today, in part, prevent a great content experience. Existing devices will favour content creation by the user, rather than consumption. This change in behaviour, even as devices improve, will remain. This change in user behaviour will also change the way content owners interact with consumers.
Subscription models	As the volume of user generated content grows, consumers will start to value editorial skills. This change in perceived value could enable the emergence of true subscription content models; the issue will be timing.
Strategic opportunities to balance major concerns:	
Mobile Web 2.0	**Content ownership, creation and DRM could hold back content owners who focus on existing business models, those who embrace user generated content, openness and web ideals will be able to progress.**

Conclusion

Some parting thoughts after a long journey ..

Businesses and their teams are under relentless pressure to deliver ever increasing performance in terms of improved profit, or margin against a backdrop of rapid change, increasing complexity, better informed competitors and demanding customers. There is an inevitable choice between short-term rewards and often survival at the expense of longer term planning and positioning.

Mobile Web 2.0 'the book' is written in a provocative style with the aim of presenting some fundamental challenges to businesses as Mobile applications becomes more central to all core business themes including revenue generation, operational efficiency, marketing opportunity, sales channels and customer care, however the impact will vary enormously by industrial sector and the activities of a company within any given value chain.

Mobile devices are challenging the way we as customers buy, consume and create. These very same devices will enable suppliers to interact with us, as customers, on a basis that they have only ever dreamed about. The same information, metadata, we are willing share with suppliers (from clickstream to attention) will improve our mobile and web experiences. Further, sharing this metadata could become a choice, with consent but without thinking, that means that we become dependent without being loyal and raising questions over who owns our digital identity. The rate of change, the increasing complexity and light footed competitors make the evolution of Mobile Web 2.0 more akin to a swamp full of alligators, sulphur geysers and quick sand, rather than a walk through a green field.

The same technology will eventually be adopted and implemented within businesses to help drive out cost and improve efficiency. The principles of Mobile Web 2.0 i.e. : Mobile content and the changing balance of power; I am not a number – I am a tag; The mobile phone network is the computer; Multilingual implications; Digital Convergence; The disruptive power of Ajax and Mobile Widgets; Location based services and Mobile Search, have highlighted the key aspects of change in the emerging new competitive market.

However, the 'customer is still king' even though the customer is becoming more difficult to service. These new mobile orientated customers want houses without walls, games without frontiers, opportunities without rules; delivered on ever increasingly complex devices that are simpler to user coupled with annually decreasing pricing; to the game theorist, a doom loop. Companies who offer services with rules and without a mobile component will loose the 'youth generation' and its children.

In the New Kingdom, the Princes are Editors, the Princesses are Simplicity, Attention is Queen and Metadata is King!

Ajit Jaokar and Tony Fish

Acknowledgements

In no specific order, but in recognition of their contribution in the following areas : broadband, search, podcasting and out of the box thinking, we would like to thank:

Bill Jones

Bill Jones has developed and led European and global converging communications, media and IT businesses for over 20 years with Fortune and Times 100 companies. He recently participated in a UK government delegation to Korea and Japan covering broadband mobile and content, and led a major strategy project on DRM super-distribution in Infotainment (including user generated content) for a leading global Mobile Operator. Mr. Jones is the CEO of Global Village which recently won a prestigious EU Award for its communities work under the EU eContent program.

Alan Patrick

Alan Patrick has worked in the telecoms, networking and IP space for 20 years, has consulted for McKinsey and Coopers & Lybrand, and has held senior positions at British Telecom, Globix Corporation and Jacobs Rimell. He is one of the founding Principals of Broadsight, a strategic and operational consultancy specialising in all aspects of the digital multimedia field.

Sam Sethi

Sam Sethi has been involved in the IT business for the last 15 years in a variety of roles including, Director at Gateway Computers, Netscape's European Marketing Manager and several technical and marketing roles at Microsoft. Sam has an MBA from the London Business School and a commission in the Parachute Regiment.

Aenil Premji

Aenil Premji has over 20 years experience in fixed/mobile communications and content services in corporate and start-ups. Aenil has held a variety of executive roles including, COO of a Music to Mobile Download start-up and VP Proposition Management at T-Mobile International where he launched their 3G business services. At Vodafone Aenil, as Director OEM Product & Alliances, defined and launched the 'Connected by Vodafone' partner solution and sales programme and was previously Director of Marketing, Vodafone (UK) Multimedia.

About The Authors

Ajit Jaokar

Ajit is an innovator and a pioneer in the Mobile Data Industry. He is a member of the Web 2.0 workgroup (http://www.web20workgroup.com) and also chairs Oxford University's Next generation Mobile applications panel (www. forumoxford.com). Recent speaking/media appearances have included CNN Money, Ajax World (New York) and Mobile Phone conference (2006) (Seoul).

Currently, Ajit plays an advisory role to a number of Mobile start-ups in the UK and Scandinavia. He also works with the government and trade missions of countries in an advisory capacity including South Korea and Ireland.

Ajit lives in London, UK but has three nationalities (British, Indian and New Zealander) and is proud of all three. Ajit is a member of the RSA (The Royal Society for the Encouragement of Arts, Manufactures and Commerce - www. thersa.org)

He is a prolific blogger at www.opengardensblog.futuretext.com. Ajit also manages the Mobile Web 2.0 spotlight at www.sys-con.com. He can be contacted at ajit.jaokar@futuretext.com .

Tony Fish

Tony has been involved for over 18 years in the mobile, wireless, telecom and satellite industries and has been associated with hi-tech investing since his first IPO in 1994. Tony divides his time between his non-exec roles, funding opportunities, board advisory work and speaking. Tony is known for his probing questioning, clear decision making, simple, no-nonsense attitude and robust financial views and controls. Tony has an excellent grasp of strategic and economic issues relating to TMT businesses, their growth and survival, and brings an innovative flare to deal making, executing merger and acquisitions and fund raising. Tony is an experienced key note speaker and chairman and has addressed numerous organisations on TMT trends and global issues. He has particular expertise in the development of innovation through partnership.

Chairman:-Liquid Lists.

Non-exec Director for:-2Ergo, Chronos Technology, JVTV (Subtv & DOT Mobile) and SAS.

Advisory Board member:-Probability Games, Inventive Capital.

Founder:-AMF Ventures.

Tony worked for O2 where he founded their third party developer community and £100m investment fund, anfactory a $550m investment fund as Global Head of Telecoms Investing, BT as a Director of Mergers and Acquisitions in internet and multimedia, Intercai Mondiale as a Principal for multimedia and GEC Marconi where he helped launch VideoPhones in 1990.

Tony has an MBA from Bradford School of Management; a BEng from Reading University, is a Chartered Engineer, a Fellow of the Institution of Engineering and Technology (FIET) and is a member of the Chartered Institute of Marketing (MCIM).

Tony is based in London and can be contacted at tony.fish@amfventures.com. Tonys blog is at www.mobileweb20.com.

Index

Ajit Jaokar and Tony Fish